Maturation Phenomenon in Cerebral Ischemia II

Springer
Berlin
Heidelberg
New York
Barcelona
Budapest
Hong Kong
London
Milan
Paris
Santa Clara
Singapore
Tokyo

U. Ito T. Kirino T. Kuroiwa I. Klatzo (Eds.)

Maturation Phenomenon in Cerebral Ischemia II

Neuronal Recovery and Plasticity

With 53 Figures and 14 Tables

Springer

Umeo Ito
Musashino Red Cross Hospital
Department of Neurosurgery
1-26-1, Kyonan-cho, Musashino-shi
180 Tokyo
Japan

Takaaki Kirino
University of Tokyo
Faculty of Medicine
Department of Neurosurgery
7-3-1, Hongo, Bunkyo-ku
113 Tokyo
Japan

Toshihiko Kuroiwa
Tokyo Medical and Dental University
Department of Neuropathology
Medical Research Institute
Yushima 1-5-45, Bunkyo-ku
113 Tokyo
Japan

Igor Klatzo
National Institutes of Health
Former Chief of the
Laboratory of Neuropathology
and Neuroanatomical Sciences
Bethesda, MD 20892-4128
USA

ISBN-13:978-3-540-61673-3 e-ISBN-13:978-3-642-60546-8
DOI: 10.1007/978-3-642-60546-8

Library of Congress Cataloging-in-Publication Data. Maturation phenomenon in cerebral ischemia: neuronal recovery and plasticity/[edited by] U. Ito . . . [et al.]. p. cm. Proceedings of a symposium held in Tokyo, April 1996. Includes bibliographical references and index. ISBN-13:978-3-540-61673-3 1. Cerebral ischemia – Pathophysiology – Congresses. 2. Neurons – Congresses. 3. Apoptosis – Congresses. I. Ito, U. (Umeo). RC388.5.M363 1997 616.8′1 – dc20 96-36379

Cover design: Design & Production GmbH, Heidelberg

Typesetting: Best-set Typesetter Ltd., Hong Kong

SPIN: 10528482 19/3133/SPS – 5 4 3 2 1 0 – Printed on acid-free paper

Preface

Maturation phenomenon, first described by Ito et al. (1975), are the delayed postischemic changes that develop after an ischemic insult. The time length of the delay differs in various parameters of ischemic injury. A general law is that the length of delay in the development of post-ischemic changes is directly related to the intensity of the ischemic insult. Delayed neuronal death of CA1 pyramidal cells in the hippocampus, which was precisely described by Kirino (1982), is a classic example of a maturation phenomenon and has led to many investigations in this area. It has become increasingly apparent that delayed ischemic changes of a chronic character may develop in many other neuronal locations outside the hippocampus, and even in areas not affected directly by ischemia.

In September 1990, we organized the first international symposium on the "Maturation Phenomenon in Cerebral Ischemia" as the satellite symposium of the eleventh International Congress of Neuropathology in Kyoto. At this symposium, we discussed the nature and basic features of maturation phenomenon. Proceedings of the meeting were published by Springer-Verlag.

Since the first international symposium on the maturation phenomenon, new developments, particularly in the field of molecular biology, have been rapidly providing information about the molecular nature and dynamics of the mechanisms concerned with cell death and recovery. It can be assumed that maturation phenomenon represents a continuous struggle between induced mechanisms of death and the activation of defense mechanisms leading to neuronal recovery. During the past decade, investigations into ischemic brain injury have been focused mainly on selective neuronal death rather than on the development of focal infarction. In view of the clinical significance of the latter, studies on the developmental aspects, threshold values, and therapeutic approaches for the prevention or amelioration of cerebral infarction have become urgently needed. Elucidation of these mechanisms can open up opportunities for their therapeutic manipulation during a long-lasting "window" of the maturation phenomenon.

The second international workshop on the "Maturation Phenomenon in Cerebral Ischemia – Neuronal Recovery and Plasticity" provided an opportunity to evaluate and discuss various factors and mechanisms involved in ischemic neuronal injury, recovery, and plasticity. The last hour of this workshop was devoted to a round table discussion to establish the main areas of progress and promising avenues for future investigation.

This meeting was also held to commemorate the outstanding scientific work of Dr. Igor Klatzo, a career which has spanned more than 38 years at the National Institutes of Health in the USA.

May 1996 UMEO ITO and coeditors

References

Ito U, Spatz M, Walker J Jr, Klatzo I (1975) Experimental cerebral ischemia in mongolian gerbils. I. Light microscopic observations. Acta Neuropathol (Berl) 32: 209–223

Kirino T (1982) Delayed neuronal death in the gerbil hippocampus following ischemia. Brain Res 239: 57–69

Contents

MATURATION PHENOMENON IN CEREBRAL ISCHEMIA I
–NEURONAL RECOVERY AND PLASTICITY–

List of First-Named Authors

BAETHMANN, A.
Institute of Surgical Research, Klinikum Großhadern, 81366 München, Germany

BAZAN, N.G.
Louisiana State University Medical Center School of Medicine, 2020 Gravier Street, Suite B, New Orleans, LA 70112, USA

COLBOURNE, F.
Department of Neuroscience, Faculty of Medicine, University of Calgary, 3330 Hospital Drive N.W., Calgary AB, Canada 2TN 4N1

DIEMER, N.H.
Institute of Neuropathology, University of Copenhagen, Copenhagen, Denmark

HOSSMANN, K.-A.
Max-Planck-Institute for Neurological Research, Department of Experimental Neurology, Gleueler Straße 50, 50931 Cologne, Germany

IDE, T.
Department of Neurosurgery, Faculty of Medicine, University of Tokyo, 7-3-1 Hongo, Bunkyo-ku, Tokyo 113, Japan

IKAWA, Y.
Department of Retroviral Regulation, Tokyo Medical and Dental University, Medical Research Division, Yushima, Bunkyo-ku, Tokyo 113, Japan

IKEDA, J.
Department of Physiological Chemistry, Graduate School, Tokyo Medical and Dental University, 1-5-45 Yushima, Bunkyo-ku, Tokyo 113, Japan

ITO, U.
Department of Neurosurgery, Musashino Red Cross Hospital, 1-26-1 Kyonan-cho, Musashino-shi, Tokyo 180, Japan

JOHANSSON, B.B.
Department of Neurology, Lund University Hospital, 22185 Lund, Sweden

Kawahara, N.
Stroke Branch, National Institutes of Health, Building 36, Room 4A-03, Convent Drive, MSC 4128, Bethesda, MD 20892-4128, USA

Kawai, K.
Department of Neurosurgery, Teikyo University School of Medicine, 2-11-1 Kaga, Itabashi, Tokyo 173, Japan

Kempski, O.
Institute for Neurosurgical Pathophysiology, Johannes-Gutenberg-University Mainz, 55101 Mainz, Germany

Klein, H.C.
Department of Biological Psychiatry, University Hospital Groningen, Hanseplein 1, P.O.B. 30001, Groningen 9700 RB, The Netherlands

Kogure, K.
Foundation for Brain Function and Diseases, Institute of Neuropathology, 1216 Oaza Nakase, Fukaya-city, Saitama Prefect. 366, Japan

Krieglstein, J.
Institut für Pharmakologie und Toxikologie, Fachbereich Pharmazie und Lebensmittelchemie, Phillipps-Universität, Ketzerbach 63, 35032 Marburg, Germany

Kuroiwa, T.
Department of Neuropathology, Medical Research Institute, Tokyo Medical and Dental University, Yushima 1-5-45, Bunkyo-ku, Tokyo 113, Japan

Matsumoto, M.
Division of Stroke Research, First Department of Medicine and Department of Neurology, Osaka University Medical School, 2-2 Yamada-oka, Suita, Osaka 565, Japan

Mies, G.
Max-Planck-Institute for Neurological Research, Department of Experimental Neurology, Glueler Straße 50, 50931 Cologne, Germany

Nitsch, C.
Institute of Anatomy, University of Basel, Pestalozzistraße 20, 4056 Basle, Switzerland

Shamloo, M.
Laboratory for Experimental Brain Research, Wallenberg Neuroscience Center, Lund University Hospital, 22185 Lund, Sweden

Sharp, F.R.
Department of Neurology, University of California at San Francisco, Department of Veterans Affairs Medical Center, 4150 Clement Street, San Francisco, CA 94121, USA

SPATZ, M.
Stroke Branch, National Institute of Neurological Disorders and Stroke, National Institutes of Health, 36 Convent Drive, MSC 4128, Bethesda, MD 20892, USA

TAMURA, A.
Department of Neurosurgery, Teikyo University School of Medicine, 2-11-1 Kaga, Itabashi-ku, 173 Tokyo, Japan

TOMITA, M.
Department of Neurology, School of Medicine, Keio University, 35 Shinanomachi, Shinjuku-ku, Tokyo 160, Japan

YAMADA, K.
Department of Neurosurgery, Nagoya City University Medical School, 1 Kawasumi, Mizuho-ku, Nagoya 467, Japan

I. Role of Genetic Expression – Changes in Maturation Phenomenon

Molecular Approaches to Neuronal Functions and Neuronal Cell Death*

Y. Ikawa[1], T. Yoshida[1], and N. Oyaizu[1]

Summary

This is a brief overview of some of the technologies developed and findings obtained in recent years in laboratories headed by the present author. It is hoped that some of them may contribute to a molecular understanding of neuronal functions and neuronal cell death, and also to the development of new clinical methods for rescuing brain damage lesions.

Some of our data on molecular and cellular biology of neuronal functions and neuronal cell death are reviewed here in the hope of offering some hints to neurosurgeons.

Introduction

Molecular approaches to neuronal functions have been made extensively in studies on molecular events following the gating of calcium ion channels by depolarization. The Cambrian explosion in the evolution of events is known to occur after the introduction of calcium ion usage in the living organism.

A number of transcriptional regulators have been described to immediately activate early genes and later responsive genes in the neuronal cells after neuronal stimuli.

In this paper the author describes some of the novel molecular approaches he and his colleagues have adopted in order to understand neuronal functions and neuronal systems. Hopefully, some of them may suggest a new method or provide new hints to rescue ischemic brain damage.

Results and Discussion

Krev-1 – A Dominant Tumor Suppressor Gene

The principal idea of isolating this gene originated from the famous experiment by George Klein and Henry Harris. When a HeLa cell was fused with a normal human

---

* Presented at the 2nd International Workshop on Maturation Phenomenon in Cerebral Ischemia – Neuronal Recovery and Plasticity, April 1, 1996, Tokyo, Japan

[1] Dept. of Retroviral Regulation, Tokyo Medical and Dental University Medical Research Division, Yushima, Bunkyo-ku, Tokyo 113, Japan

Maturation Phenomenon in Cerebral Ischemia II
U. Ito et al. (Eds.)

fibroblast, cancer phenotypes disappeared, and the hybrid cells were non-tumorigenic in athymic nude mice. When a particular chromosome was deleted in those cells, they recovered their tumorigenicity, suggesting the presence of a dominantly functioning tumor-suppressor gene in that chromosome.

Expanding on this idea, Noda and others [6] created a cDNA expression library of normal human fibroblasts and transfect this library into cells of the DT cell line, a NIH3T3 cell line transformed by K-*ras*; they isolated several flat revertants which lost most of their neoplastic phenotypes.

From such a revertant the K*rev*-1 gene, also called *rap*1A, was isolated [5]. Although this unique *ras*-homologous gene has been conserved from nematode to man, the actual biological function of the gene is still unknown, except for its high expression in the neuronal tissues. Two amino acids, the 30th and 31st amino acids (E K D E) from the N terminus, upstream of the effector domain, differentiate its tumor-suppressing function from its tumor-inducing *ras* function [4].

Urayama et al. [8] observed the increase of K*rev*-1 expression in the dorsal horn of the rat spinal cord in response to pain stimulus (Capsaicin, etc). Extensive distribution of K*rev*-1 in the neuronal tissues, including the central nervous system (CNS), and possible presence of distinct effectors between *ras* and K*rev*-1 G proteins, may open a new direction in neurobiology.

Highly Efficient Embryonal Technology

The author's group recently developed a new embryonal stem (ES) cell line (TT2) from a 4-day F_1 embryo between C57BL/6 and CBA/2 mice [9].

When TT2 cells, gene-manipulated or unmanipulated, were microinjected into eight cell-stage fertilized eggs of white, inbred-mouse origin, those ES cells immediately migrated to the center of the core cell mass of the blastocysts. Implantation of the microinjected embryos into pseudopregnant white female mice resulted in the highly efficient development of TT2-originated agouti-colored offspring (Table 1).

Yagi and his colleagues [10] also developed a new targeting vector using the diphtheria toxin A gene as a second selection marker. These systems are quite useful in the generation of gene knock-out mice and are now commercially available from the Life Tech. Oriental Co. Ltd., Tokyo 103, Japan.

Table 1. Mutated TT2 cells: germline-differentiating potency

Clone name	Gene locus	No. of males mated and fertile	No. of males generating TT2-derived offspring at			
			100%	100%–50%	50%–0%	0%
$F_2$45	fyn	8	8	0	0	0
C2-S	csk	6	6	0	0	0
T18	p53	18	16	0	1	1
T48	p53	12	11	0	0	1
F18	Tenascin	18	18	0	0	0

Table 2. Success of 7-day-old pups to reach the nipples of anesthesized mothers in 10 min

7-day-old pups	Activated mammary gland	Unactivated mammary gland
Fyn (+/–)	16/28	10/28
Fyn (–/–)	30/58	0/58[a]

[a] Two reached nipples in 15 min.

Using the above embryological technologies, a non-receptor type tyrosine kinase, Fyn, was knocked out; Fyn-deficient pups from Fyn-deficient parents had difficulty in finding the inactivated nipples and died several days after birth [11]. There was also deformity in their olfactory bulb and in the CA3 region of the hippocampus. When a healthy pup was provided to activate the mother's nipples, however, those *fyn*-deficient pups survived. Seven-day-old Fyn-deficient pups were placed beside their anesthetized mother, with or without activated nipples, and the above observation was confirmed (Table 2).

Application of Comparative Genetics

A nematode consists of 1090 cells and 131 of these die as a result of apoptotic changes; about one third of the cells in these organisms are related to neuronal function. Many uncoordinated movement mutants (Unc-mutants) have been isolated. Dr. K. Ando, my former colleague at the Tsukuba Life Science Center, RIKEN, cloned and characterized a mouse homologue of the Unc-18 gene. The Unc-18 nematode shows acetylcholine (ACh) retention during development and is resistant to Trichlorphon, an ACh-esterase inhibitor. The Unc-18 gene was found to be homologous to Sec-1 in yeast [1]. There are two types of mouse Unc-18 homologue: brain-specific *synap*-1 (n-*sec*-1) and ubiquitously expressed *synap*-2. Mouse *synap*-1 restores dysfunction of ACh- and γ-aminobutyric acid (GABA-) release in the nematode Unc-18 mutants, and also restored hyper-sensitivity to levamisole, an ACh-receptor agonist (Fig. 1) [3].

This kind of comparative genetic method may be useful in investigations of genes involved in neuronal functions.

Forced Induction of Apoptosis

Yoshida et al. [12] recently constructed a chimeric gene for the extracellular human Fas and intracytoplasmic TNFR-1 receptors. By applying an agonistic monoclonal antibody, CH-11, they were able to induce cell death in several cell lineages after introduction of the gene (Fig. 2).

The cytoplasmic domains of Fas and TNFR-1 possess distinct death domains which are found by different cell death-mediating proteins; changing the cytoplasmic genes of the receptors enables the investigation of subtle differences in the death signals in induced apoptosis and necrosis in various cell lineages.

Fig. 1. Sensitivity to levamisole

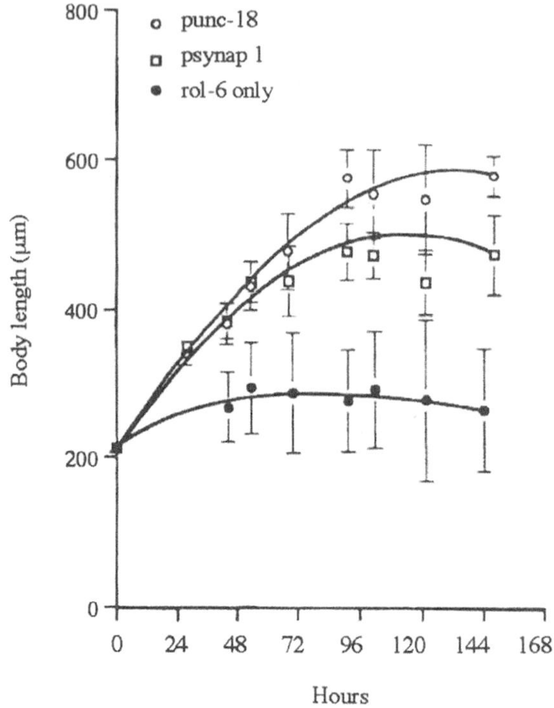

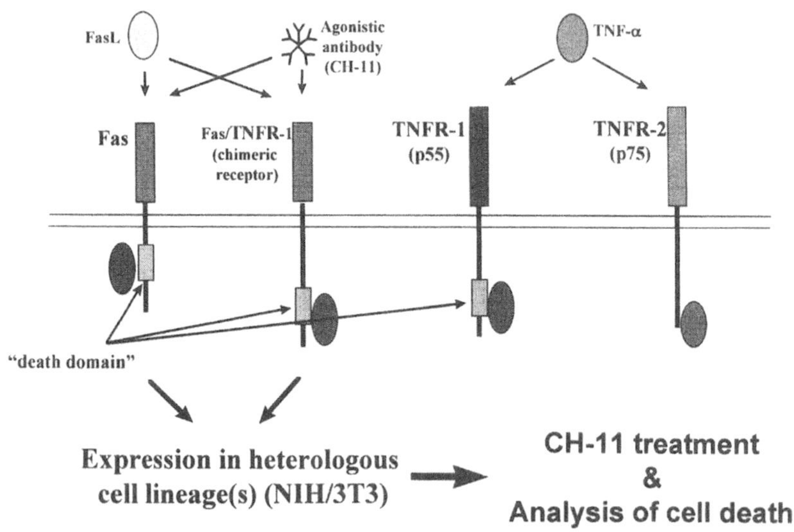

Fig. 2. Chimeric Fas/TNFR-1 receptor in heterologous cells activated by CH-11 treatment. *TNF-α*, tumour necrosis factor-*α*; *TNFR-1*, tumour necrosis factor receptor-1

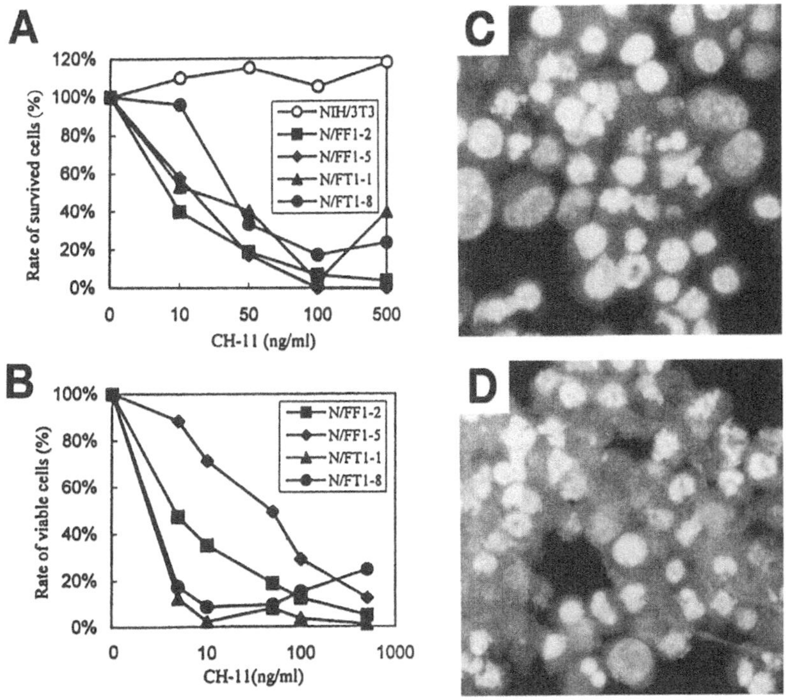

Fig. 3A–D. Apoptosis induced in Fas/TNFR-1-introduced NIH3T3 cells augmented by CH-11. **A** Rate of survival cells. **B** Rate of viable cells. **C** Fas/Fas-NIH3T3 cells 24 h after CH-11 treatment. **D** Fas/TNFR-1-NIH3T3 cells 24 h after Ch-11 treatment

When this chimeric gene was introduced to NIH3T3 cells, cell death was induced within 24 h with less distinct nuclear particulization and DNA fragmentation (Fig. 3). Several cell death-related pathways can be studied in the neuronal cells using such experimental systems, and some of their inhibitors may prevent neuronal cell damage in the brain.

Gene Transfer to Neuronal Cells

Gene transfer to damaged cells or damaged tissues in the brain may be one target of future projects in brain surgery. The genes to be transferred are those that block apoptotic or necrotic changes or neurotrophic factors (Fig. 4).

Vectors have been developed from adenoviruses, and retroviruses. Since most neuronal cells do not replicate, human immunodeficiency virus (HIV) may be the best source for the development of the vector because the HIV genome possesses the machinery for integration into the nucleus of a resting cell.

One such HIV vector has been developed by Shimada and Miyake [7] in which viral proteins are separately produced using a cytomegalovirus promoter and the vector plasmid contains the neomycin resistant gene (neo^r) for selection (Fig. 5). A means of

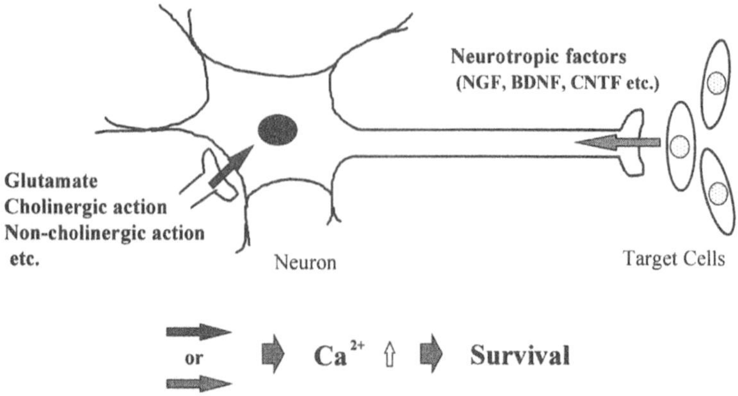

Fig. 4. The survival of neurons. Genes of neurotrophic factors to be transferred to damaged neuronal cells. *NGF*, nerve growth factor; *BDNF*, brain-derived neurotrophic factor; *CNTF*, ciliarg neurotrophic factor

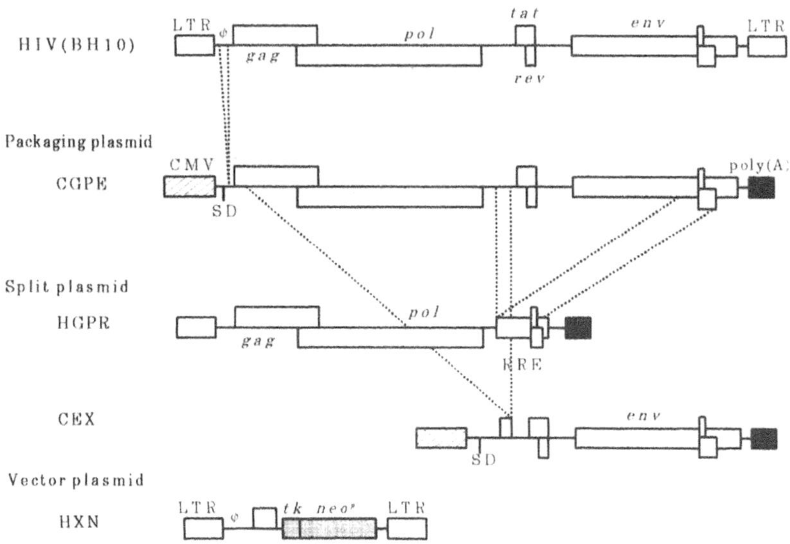

Fig. 5. Human immunodeficiency virus (HIV)-derived vector for resting cells. To ensure safety of the vector, viral components are produced by separate plasmids. *CMV*, cytomegalovirus; *LTR*, long terminal repeat; RRE, rev-responsive element; *SD*, splice donor

transferring some of the HIV genes to a Moloney murine leukemia virus (Mo-MuLV) vector is also being designed to develop a vector for resting cells.

Establishment of Neuronal Progenitor Cell Lines

It has long been believed that neuronal cells do not replicate in the adult animal and are weak in the anoxic state. However, this is not the case with neuronal cells in the hippocampal and olfactory regions. Gage et al. [2] were successful in establishing a neuronal progenitor cell line from the dentate gyrus region of a rat using basic fibroblast growth factor (b-FGF). These cells have histochemical and ultrastructural characteristics of neurons, and after introduction of the marker LacZ gene, they were transplanted and were observed to migrate to replace the damaged dentate gyrus lesion. These cells have now been maintained in vitro for over a year.

The neuronal stem cells appear to be quite resistant to the anoxic state and there is some hope of developing human neuronal progenitor cell lines from different parts of the brain. These cell lines might be useful for the replacement of the damaged lesions of the brain.

Ono et al. (personal communication) were successful in the in vitro maintenance of cerebellar granule neurons under depolarization conditions in which they observed suppression of apoptosis in association with parathyroid hormone-related protein expression.

Acknowledgments. The author gratefully acknowledges the organizers, especially Dr. I. Klatzo, and Dr. U. Ito, for their invitation to speak at this 2nd International Workshop on Maturation Phenomenon in Cerebral Ischemia, 1996. He is also thankful to Ms. J. Clarkin and Mr. K. Aizaki for their help in preparing the manuscript.

References

1. Aalto MK, Keranen S, Ronne H (1992) A family of proteins involved in intracellular transport. Cell 68: 181–182
2. Gage FH, Coates PW, Palmer TD, Kuhn HG, Fisher LJ, Suhonen JA, Peterson DA, Suhr ST, Ray J (1995) Survival and differentiation of adult neuronal progenitor cells transplanted to the adult brain. Proc Natl Acad Sci USA 92: 11879
3. Gengyo-Ando K, Kitayama H, Mukaida M, Ikawa Y (1996) A murine neural specific homologue corrects cholinergic defects in C. elegans Unc-18 mutants. J Neurosci 16: 6695–6702
4. Kitayama H, Noda M (1993) *Krev*-1 and the related genes. In: Lacal JC, McCormick F (eds) The ras super family of GTPases. CRC, Boca Raton, p 231
5. Kitayama H, Sugimoto Y, Matsuzaki T, Ikawa Y, Noda M (1989) A *ras*-related gene with transformation suppressor activity. Cell 56: 77
6. Noda M, Kitayama H, Matsuzaki T, Sugimoto Y, Okayama H, Bassin RH, Ikawa Y (1989) Detection of genes with a potential for suppressing the transformed phenotype associated with activated *ras* genes. Proc Natl Acad Sci USA 86: 162
7. Shimada T, Miyake K (1996) Gene transfer into non-dividing cells by means of HIV-based vectors. Proceedings of the 2nd Annual Meeting of the Japan Society of Gene Therapy, p 107
8. Urayama O, Murakoshi T, Ikawa Y (1995) *Krev*-1 protein is abundantly expressed in the rat spinal cord. Biochim Biophys Acta 1243: 446–452
9. Yai T, Tokunaga T, Furuta Y, Nada S, Yoshida M, Tsukada T, Saga Y, Takeda N, Ikawa Y, Aizawa S (1993) A novel ES cell line. TT2, with high germline-differentiating potency. Anal Biochem 214: 70–76
10. Yagi T, Nada S, Watanabe N, Tamemoto H, Kohmura N, Ikawa Y, Aizawa S (1993) A novel negative selection for homologous recombinants using diphtheria toxin A fragment gene. Anal Biochem 214: 77–86
11. Yagi T, Aizawa S, Tokunaga T, Shigetani Y, Takeda N, Ikawa Y (1993) A role for Fyn tyrosine kinase in the suckling behavior of neonatal mice. Nature 366: 742–745
12. Yoshida T, Sato A, Ikawa Y (1996) Both human fas and chimeric fas/TNFR-1 mediate complete cell death in murine NIH/3T3 cells with less distinct nuclear fragmentation. Proc Jpn Acad 72: 16–21

Evolution of Gene Expression in Hippocampus Following Focal and Global Ischemia

F.R. Sharp[1], J. Honkaneimi[2], B. States[1], and S.M. Massa[1]

Summary

Permanent middle cerebral artery (MCA) occlusions were produced using the suture model in adult Sprague Dawley rats. Heat shock protein (HSP)70 immunostaining and terminal deoxynucleotide transferase uridine nick end labeling (TUNEL) staining were performed 1, 3, and 7 days later. Half of the animals at 1 and 3 days had HSP70- and/or TUNEL-stained neurons in the CA1 sector of hippocampus. Double labeling on the same section showed that HSP70-stained neurons were generally TUNEL negative, and TUNEL-stained neurons were generally HSP70 negative.

Examination of hippocampus in gerbils following global ischemia showed that TUNEL positive CA1 neurons were detected in CA1 at 3 days following 5 min ischemia, and at 1 day following 10 min ischemia. Oligonucleotides were used to show that bcl-2 and bcl-x (all forms) mRNAs were induced on Northern blots of hippocampus at 1 day following 5 min and 10 min of ischemia. In situ hybridization showed that the bcl-2 and bcl-x mRNAs were expressed in the CA1 neurons destined to die.

The data shows that CA1 neurons that express HSP70 protein at 1–3 days following either focal or global ischemia are probably "reversibly injured" and survive. Longer durations of ischemia accelerate the rate at which CA1 neurons become TUNEL positive. The induction of apoptosis-related genes in CA1 neurons destined to die, combined with the TUNEL staining and DNA laddering data, supports the concept that apoptosis-related mechanisms can be implicated in the delayed death of CA1 neurons following focal and global ischemia.

Introduction

Much has been learned about the mechanims of neuronal cell death in general. However, the phenomenon of delayed cell death described by Kirino [7] that occurs in CA1 neurons following brief durations of global ischemia has eluded definitive explanation. The finding that KA/AMPA glutatmate antagonists best protect CA1 hippocampal neurons against global ischemia [14], and the finding that the subunits of the

[1] Department of Neurology, University of California at San Francisco, Department of Veterans Affairs Medical Center, 4150 Clement Street, San Francisco, CA 94121, USA
[2] Department of Neurology, University of Tampere, Tampere, Finland

Maturation Phenomenon in Cerebral Ischemia II
U. Ito et al. (Eds.)

KA/AMPA receptors may change so that they begin to flux calcium [13] may help explain how the initial ischemic event sets the cell death machinery in motion. However, even these important observations have not satisfactorily explained why the process is so delayed.

A number of recent studies have demonstrated that DNA fragmentation occurs in the hippocampus, and in particular in CA1 neurons [5, 8]. This has been demonstrated using TUNEL staining which demonstrates cells with fragmented DNA; moreover, evidence of DNA fragmentation has been demonstrated using DNA laddering, which appears to be more specific for the process of apoptosis and actions of endonucleases [10, 11]. Though an apoptotic type of cell death might help explain CA1 neuronal demise, there has been little other data suggesting an important role for apoptosis in cerebral ischemia.

The present study was aimed at examining two issues: What is the relationship between TUNEL-stained, presumably apoptotic neurons and neurons that express HSP70 protein in the hippocampus following ischemia? Secondly, are apoptosis-related genes induced in hippocampal neurons known to die following ischemia?

Materials and Methods

The middle cerebral artery (MCA) of male Sprague-Dawley rats (Bantin and Kingman; 280–320 g) was permanently occluded using the endovascular internal carotid artery (ICA) suture method of Longa et al. [9], with minor modifications. Rats were intubated and anesthesia maintained with mechanical ventilation. The right common carotid artery (CCA) was exposed at its bifurcation through a midline cervical incision. The external carotid artery (ECA) was transected and a 3-0 nylon monofilament suture, its tip rounded by heating, was inserted into the ECA stump. It was advanced into the ICA for a distance of 22 mm beyond the CCA bifurcation to occlude the origins of the MCA and proximal anterior cerebral artery. The suture was secured in place with a ligature. The animals were allowed to survive for 1 ($n = 6$), 3 ($n = 12$) or 7 days ($n = 7$) with food and water available ad libitum. Seizures were not observed in the current experiments at any time following the MCA occlusions.

Brain temperature, measured using the temporalis muscle, was maintained in the normal range before, during and for ten min after ischemia using a thermistor-controlled heating blanket. The post-ischemia hyperthermia which occurs with the suture model was not controlled. However, since the body temperature never goes over 39.5°C following MCA occlusions, this would not be high enough to induce heat shock protein (HSP)70 by itself. In addition, hyperthermia only induces HSP70 in endothelial cells and glial cells, so that the HSP70 induction in neurons in this study cannot be accounted for by hyperthermia alone. Following the MCA occlusions there was no mortality in the 1-day group (0/6, 6 survivors), 33% mortality in the 3-day group (6/18, 12 survivors), and 36% mortality in the 7-day group (4/11, 7 survivors). The brains of the animals that died showed swelling of the infarcted hemisphere.

Controls ($n = 4$) were subjected to an identical sham surgery. Controls were sacrificed 3 days following surgery since this was the time of maximal HSP70 induction in the MCA occlusion groups. At the appropriate time following surgery, ischemic animals and control animals were given a lethal overdose of chloral hydrate. Once deeply anesthetized, they were perfused through the ascending aorta with 200cc of 10% glycerol in phosphate-buffered solution (PBS). After perfusion the brains were rapidly removed and frozen in 2-methyl butane at −25°C. Brains were embedded in Lipshaw matrix, wrapped in foil and kept at −70°C until sectioning was performed.

A modification of the terminal deoxynucleotide transferase dUTP-biotin nick end-labeling technique (TUNEL) was used [3, 4, 10, 11]. Coronal 20 μm sections were cut in a cryostat at −20°C and collected on slides. Sections were fixed for 10 min in 4% paraformaldehyde and placed in 0.1 *M* PBS (pH 7.4) 2 × 15 min, 0.3% triton X-100 in PBS 2 × 15 min, and PBS 2 × 15 min. Terminal deoxynucleotidyl transferase (TdT, 1X) buffer was applied for 15 min. A mixture of TdT (300 units/ml) and biotin 14-dATP (40 μM) in 1X TdT buffer was applied and incubated at 37°C for 60 min. Slides were washed in 2X sodium citrate/sodium succinate buffer (SSC) (pH 7.0) 2 × 15 min, 2% bovine serum albumin (BSA) in PBS, and PBS 2 × 15 min. Avidin-biotin-horseradish peroxidase solution was applied for 30 min followed by sodium acetate solution (0.175 *M*) 2 × 15 min. Staining was visualized using 0.015% diaminobenzidine (DAB), 0.001% H_2O_2 and 1% nickel sulfate in 0.175 *M* sodium acetate. After staining, slides were rinsed with water. Positively stained cells had dark blue to black, round nuclei.

After processing for DNA nick end-labeling, HSP70 heat shock protein immunocytochemistry was carried out on the same sections. HSP70 mouse monoclonal antibody C92 (1:4000 in 1% BSA) was applied to sections and incubated at 4°C overnight. Sections were washed 2 × 15 min in 1% BSA and incubated in second antibody (biotinylated sheep antimouse IgG) (1:200 in 1% BSA) for 30 min. After washes in PBS 2 × 15 min, avidin-biotin-horseradish peroxidase solution was applied as above. Sections were washed in PBS and placed in 0.015% DAB and 0.001% H_20_2 in PBS for 2–8 min. The slides were washed with water, dehydrated in ethanol and cover-slipped. HSP70-stained cells had brown staining of the cell body and dendrites. Nissl staining was performed on adjacent sections. The C92 monoclonal antibody to HSP70 was raised to human HSP70 protein. It detects two bands on Western blots of ischemic gerbil brain and it detects the HSP70 protein expressed from the rat HSP70 cDNA [6, 9, 12].

Adult male gerbils were anesthetized with methoxyflurane. Global ischemia was produced by occluding both carotid arteries for a period of 5 or 10 min. Shams had identical surgeries except that the carotids were not occluded. Body temperature was maintained in the range of 37.5°C ± 5°C with a heating lamp. Subjects were allowed to survive 6 h, 24 h, or 72 h following surgery.

At the appropriate time following the surgery subjects were anesthetized with methoxyflurane and decapitated. Brains were taken for either in situ hybridization or Northern blots. For in situ hybridization, brains were removed and frozen. Frozen sections were cut in a cryostat and mounted onto probe on slides. At least four sections from each animal were hybridized with each oligonucleotide probe. The oligonucleotides were chosen from regions of each gene present in all known splice variants of each gene showing the highest degree of homology among different species.

The oligonucleotides were labeled with ^{35}S-dATP using terminal deoxynucleotidyl transferase. The sections were air dried at room temperature and hybridized at 42°C for 18 h in a mixture of 4 × SSC, 50% formamide, 1× Denhardts, 1% sarcosyl, 0.02 *M* phosphate buffer (pH 7), 10% dextran sulfate, 500 μg/ml heat denatured salmon sperm DNA, 200 m*M* dithiothreitol and 1 × 10^7 cpm/ml of the labeled probe. After hybridization, the sections were washed four times for 15 min each in 1 × SSC at 55°C and thereafter left to cool for 1–3 h. Sections were then dipped in distilled water, 75% ethanol and air dried. They were then autoradiographed using Kodak SB5 film.

For Northern blots hippocampi were disected from 32 control and 32 ischemic gerbils 24 h following 5 min of global ischemia. Three separate samples were processed for Northern blots. The tissue was homogenized in lysis buffer, incubated in oligo dT cellulose and poly(A)RNA isolated using the Fast Tract Isolation kit (Invitrogen, San Diego, CA). To remove contaminating DNA, the samples were incubated with 5 U/ml of DNase and 200 U/ml of RNAsin at 37°C for 30 min. The RNA was precipitated in ethanol, dissolved in water and the optical density (OD) determined. Four micrograms of RNA was electrophoresed in 1.5% agarose containing 1 × (*N*-2-morpholino) propane sulfonic acid (MOPS) and 7% paraformaldehyde, and transferred to a nylon membrane. Three such membranes were prepared for the ischemic group and three for the sham group. The RNA was cross linked to the membranes with UV light at 254 nm. The membrane was stained with 0.02% methylene blue and the position of the 18S- and 28S-bands marked. Prehybridization was conducted for 1 h at 42°C with 6XSSC, 0.1% sodium dodecyl sulphate (SDS), 10X Denhardts, and 50 μg/ml heat denatured salmon sperm DNA. The oligonucleotide probes for bcl-2, bcl-x, and interleukin-1β-converting enzyme (ICE) were labeled using TdT with [^{32}P]dATP. The membranes were hybridized at 37°C overnight in 6XSSC, 1% SDS, and 1–4 × 106 cpm/ml of labeled probe. To verify the relative amounts of mRNA transferred, the membranes were hybridized with [^{32}P]dATP labeled oligonucleotide probe that recognized cyclophilin. After hybridization the membranes were washed to a stringency of 6XSSC, 0.1% SDS at 55°C. The membranes were then autoradiographed.

Results

There was no DNA nick end-labeling, evidence of infarction, or any edema in any brain region of sham operated control subjects. In the brains of the ischemic animals, DNA nick end-labeling resulted in labeled cell nuclei exhibiting a dark blue to black color. Labeled DNA sometimes appeared to aggregate in more densely stained masses which had the appearance of being compacted against the nuclear membrane. Other cells, however, exhibited homogeneous staining of labeled nuclei which could be consistent with either necrosis or apoptosis. No attempt was made during cell counting to differentiate between these two cell types.

DNA nick end-labeling in cortex was detected at 1 day, was much greater at 3 days, and had disappeared in cortex by 7 days. The areas of infarction, mainly in cortex and basal ganglia, had many stained cells at all time points with the amygdala and adjacent basal cortex exhibiting the greatest number. Outside the areas of infarction, DNA nick end-labeled cells were observed in the ipsilateral hippocampus in two out of six of the 1-day animals, and in five out of 12 of the 3-day animals. Hippocampal DNA fragmentation was most pronounced in the CA1 region, particularly the medial aspect. In one animal at 3 days following ischemia, the contralateral CA1 had numerous TUNEL stained cells.

In animals subjected to the permanent MCA occlusions, HSP70 protein induction within areas of infarction was confined mainly to endothelial cells which may survive the ischemia. In some animals a rim of HSP70 immunostaining occurred in dorsolateral cortex, a borderzone between the MCA and ACA territories. In these areas many but not all neurons stained intensely for HSP70 protein as did variable numbers of glial cells.

In areas remote from the infarction 50% of the animals in the 1-day group and 50% of the animals in the 3-day group expressed HSP70 in the hippocampus. HSP70 was expressed in the contralateral hippocampi of three animals. HSP70 induction in hippocampus at 1 and 3 days following ischemia was greatest in neurons in the CA1 region, but also occurred in CA3 and the dentate hilus.

At the edge of the infarcted area in dorsomedial cortex, as the MCA territory extended medially towards the anterior cerebral artery territory, the number of DNA nick end-labeled cells diminished while the number of HSP70 positive neurons increased. The border between these two territories corresponded to the area of greatest regional commingling of cells that stained positively for either HSP70 induction or DNA fragmentation. This area was located immediately adjacent to the infarction. In this area a few double-stained cells could be located on most cortical sections. The double-labeled cells in cortex represented a minority of the total population of cells.

In the cases where hippocampal DNA fragmentation was observed in most of the CA1 pyramidal neurons, CA3 pyramidal neurons and dentate hilar neurons stained intensely for HSP70. As the CA1 field coursed laterally in these animals, the number of DNA nick end-labeled cells diminished while the number of HSP70-positive cells increased. In animals that had lesser degrees of DNA fragmentation in CA1, HSP70 induction occurred in neurons at the margins of small zones of CA1 where only DNA nick end-labeled cells were detected. As in cortex, hippocampal cells that stained for HSP70 protein were observed adjacent to cells that demonstrated DNA nick end-labeling.

Baseline levels of mRNA were detected in the hippocampus of sham operated animals using all three probes to bcl-2, bcl-x, and ICE. Although only one transcript was detected for bcl-2, at least two mRNA transcripts were detected for bcl-x, and at least three bands were detected for ICE. At 24 h and 72 h following ischemia, bcl-2 and bcl-x mRNAs (both transcripts) were markedly induced on the Northerns compared to the levels seen in shams. There was no induction of any of the ICE-like transcripts that could be detected by Northern blotting at 24 h and 72 h following ischemia.

In situ hybridization showed that bcl-2 mRNA and bcl-x mRNA could be detected in the CA1 and CA3 pyramidal cell layers in the hippocampus of sham operated gerbils. At 24 h following 5 min of global ischemia, bcl-2 mRNA was markedly induced in virtually all CA1 pyramidal neurons. At the same time, bcl-x mRNA was also markedly induced in CA1 pyramidal neurons. There was no apparent induction of ICE-like mRNAs, although at 72 h a faint band of signal might have been detected around the dying neurons.

TUNEL staining showed no positive CA1 neurons at 24 h following 5 min of global ischemia. By 72 h following 5 min of global ischemia most CA1 neurons were TUNEL positive. However, at 24h following 10min of global ischemia TUNEL positive neurons were detected in medial and lateral CA1.

Discussion

The experiments demonstrate that hippocampal neurons are damaged half the time following permanent MCA occlusions in adult rats. HSP70-stained neurons, in both cortex and hippocampus, generally survive the ischemic injury since most of these neurons are TUNEL negative. It is possible to describe at least two penumbras around areas of infarction: a region very near the infarct where TUNEL positive and HSP70 positive neurons are found; and secondly, a region which extends further from the infarction where HSP70-stained neurons are mainly found. Lastly, it appears that neurons that die a delayed death in hippocampus following either focal or global ischemia generally do not express HSP70 protein.

The finding that both bcl-2 and bcl-x mRNAs are induced in CA1 neurons following global ischemia suggests that genes involved in regulating apoptotic pathways in other cells are also induced in ischemic neurons. Though the mechanism of the bcl-2 and bcl-x mRNA induction is not known, it is tempting to speculate that activation of KA/AMPA receptors during the ischemia is an early event. The induction of bcl-2 and bcl-x is obviously a much later event. Since both bcl-2 and at least bcl-xl would be expected to tend to protect the CA1 neurons [1, 2], it seems reasonable to assume that either the bcl-2 protein expression is insufficient to protect the cells, or that the overexpression of apoptosis-related killer proteins exceeds the protection afforded by bcl-2 and other protective proteins. It is also possible that chronic calcium overload of the cells results in the activation of endonucleases which are not countered by suffi-

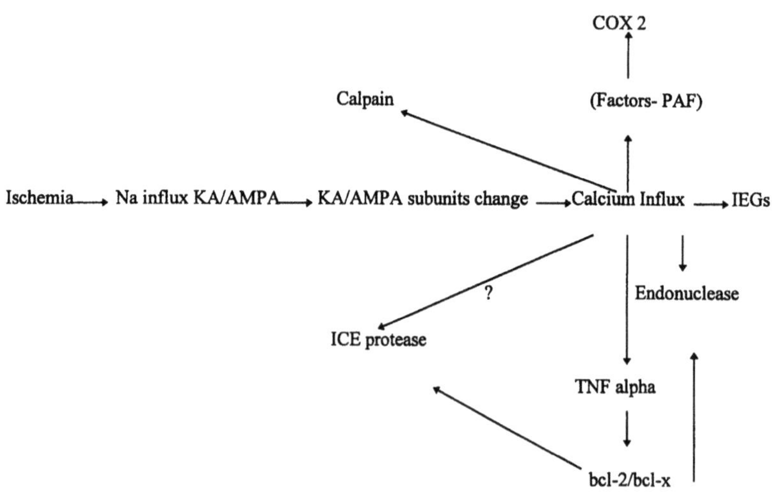

Fig. 1. Possible mechanisms of gene induction following cerebral ischemia

cient bcl-2 resulting in primary DNA degradation. These possibilities are outlined in Fig. 1.

References

1. Boise LH, Gonzalezgarcia M, Postema CE, Ding LY, Lindsten T, Turka LA, Mao XH, Nunez G, Thompson CB (1993) bcl-x, a bcl-2 related gene that functions as a dominant regulator of apoptotic cell death. Cell 74: 597–608
2. Garcia I, Martinou I, Tsujimoto Y, Martinou JC (1992) Prevention of programmed cell death of sympathetic neurons by the bcl-2 proto-oncogene. Science 258: 302–304
3. Gold R, Schmied M, Giegerich G, Breitschopf H, Hartung HP, Toyka KV, Lassmann H (1994) Differentiation between cellular apoptosis and necrosis by the combined use of in situ tailing and nick translation techniques. Lab Invest 71: 219–225
4. Honkaniemi J, Massa SM, Sharp FR (1996) Apoptosis associated genes are induced in gerbil hippocampus following global ischemia. Rest Neurol Neurosci 9: 227–230
5. Kihara S, Shiraishi T, Nakagawa S, Toda K, Tabuchi K (1994) Visualization of DNA double strand breaks in the gerbil hippocampal CA1 following transient ischemia. Neurosci Lett 175: 133–136
6. Kinouchi H, Sharp FR, Hill MP, Koistinaho J, Sagar SM, Chan PH (1993a) Induction of 70-kDa heat shock protein and hsp70 mRNA following transient focal cerebral ischemia in the rat. J Cereb Blood Flow Metab 13(1): 105–115
7. Kirino T (1982) Delayed neuronal death in the gerbil hippocampus following ischemia. Brain Res 239: 57–69
8. Li Y, Chopp M, Jiang N, Zaloga C (1995) In situ detection of DNA fragmentation after focal cerebral ischemia in mice. Brain Res Mol Brain Res 28: 164–168
9. Longa FM, Wang S, Narasimhan P, Zhang JS, Chen J, Massa SM, Sharp FR (1993) cDNA cloning and expression of stress-inducible rat hsp70 in normal and injured rat brain. J Neurosci Res 36: 325–335
10. MacManus JP, Buchan AM, Hill IE, Rasquinha I, Preston E (1993) Global ischemia can cause DNA fragmentation indicative of apoptosis in rat brain. Neurosci Lett 164: 89–92
11. MacManus JP, Hill IE, Huang ZG, Rasquinha I, Xue D, Buchan AM (1994) DNA damage consistent with apoptosis in transient focal ischaemic neocortex. Neuroreport 5: 493–496
12. Narasimhan P, Swanson RA, Sagar SM, Sharp FR (1996) Induction of the HSP70 heat shock protein and tolerance to acidosis in cultured cortical astrocytes. Glia 17: 147–159
13. Pellegrini-Giampietro DE, Zukin RS, Bennett MV, Cho S, Pulsinelli WA (1992) Switch in glutamate receptor subunit expression in CA1 subfield of hippocampus following global ischemia in rats. Proc Natl Acad Sci USA 89: 10499–10503
14. Sheardown MJ, Nielsen EO, Hansen AJ, Jacobsen P, Honore T (1990) 2,3-Dihydroxy-6-nitro-sulfamoyl-benzo(F)quinoxaline: a neuroprotectant for cerebral ischemia. Science 247: 571–574

Synaptic Messengers, Inflammatory Mediators, and Neuronal Plasticity in Cerebral Ischemia

N.G. Bazan

Summary

Cerebral ischemia activates the production of both endogenous and exogenous mediators of the inflammatory response. Using the middle cerebral artery occlusion suture model in the rat, we have found that occlusion followed by reperfusion leads to a remarkable induction of cyclo-oxygenase (COX) 2, both in the infarcted area and in the penumbra, occurring 6 h after 2 h of ischemia followed by 3 h of reperfusion. COX-1, the constitutive enzyme, does not display changes under these conditions. Recent studies in our laboratory have also shown that in a kainic acid induced brain damage model COX-2 induction occurs at the transcriptional level, since nuclear runs on assays in isolated nuclei from the hippocampus as a function of time after injury matched the profile of changes in COX-2 mRNA abundance. Moreover, platelet-activating factor (PAF) is an activator of COX-2 transcription in transfected cells, and PAF antagonists in vivo block the injury-induced COX-2 expression. Therefore phospholipase A_2 activation and PAF generation in cerebral ischemia link the formation of prostaglandins through COX-2 induction. These signaling pathways may be transneuronally controlled through excitatory amino acid neurotransmitters. It is hypothesized that phospholipase A_2 activation, PAF accumulation, and COX-2 induction are reduced during the development of enhanced tolerance to ischemic injury.

Introduction

The earliest biochemical events involving degradation of membrane phospholipids in cerebral ischemia and seizures are an accumulation of free arachidonic acids [3], eicosanoids [5], and platelet-activating factor (PAF; 1-*O*-alkyl-2-acetyl-glycero-3-phosphocholine) [20]. The stimulus for the release of these bioactive mediators is an activation of phospholipases A_2 [5] and C [1]. The release represents an overactivation of the normal physiological processes by which these lipid mediators are synthesized and exemplifies the way in which certain pools of excitable membrane phospholipids act as reservoirs for the stimulus-elicited release of lipid second messengers. One such reservoir is a small but metabolically active pool of 1-*O*-alkyl-2-arachidonoyl-glycero-

Louisiana State University Medical Center School of Medicine, LSU Eye Center and Neuroscience Center, 2020 Gravier Street, Suite B, New Orleans, LA 70112, USA

Maturation Phenomenon in Cerebral Ischemia II
U. Ito et al. (Eds.)

3-phosphocholine (alkylacyl-GPC) that upon hydrolysis by phospholipase A_2 (PLA_2) yields both arachidonic acid and lyso-PAF, a direct precursor of PAF. Alkylacyl-GPC is a substrate for a brain cytosolic PLA_2, activated and translocated to the membrane by intracellular calcium [6, 39]. Lyso-PAF is transacetylated to yield biologically active PAF via the remodeling pathway of PAF synthesis. There are at least two alternative pathways by which PAF can be generated: via the de novo synthesis route and via a coenzyme A independent transacylase. While the contribution of these pathways to PAF synthesis during neuronal stimulation has yet to be fully evaluated, the rise in intracellular calcium which stimulates the remodeling pathway also inhibits the phosphocholine transferase required in the de novo route [9].

There are high-affinity PAF binding sites in the brain [22] which are kinetically and pharmacologically distinct in synaptosomes and microsomes. The two sites differ in their sensitivity to various PAF antagonists. The *Ginkgo biloba*, a tree-derived BN52021, binds preferentially to the synaptosomal site [22], and the synthetic hetrazepine BN50730 shows specificity for the microsomal site. The synaptosomal site appears to correspond to the cloned PAF receptor [15, 21, 35] of the seven membrane spanning domain, G protein coupled receptor superfamily. The identity of the BN50730 sensitive receptor requires elucidation, but it has a specific role in PAF-induced primary-early gene expression in brain in that it differs at least in its signal transduction coupling from the synaptosomal receptor.

While the pathophysiological effects of PAF on inflammatory and immune responses in many tissues and organs have been well documented (for review see [32]), it is now becoming apparent that, via its modulation of glutamate release, PAF can also have profound effects on excitatory neurotransmission, excitotoxic neuronal damage, and synaptic plasticity.

PAF Activates Gene Transcription of the Inducible Prostaglandin Synthase and of Other Early-Response Genes

Physiological and pathological events such as NMDA receptor activation, long-term potentiation, ischemia, and seizures initiate transcription of genes encoding transcription factors and thus have the potential to initiate cascades of gene expression [8, 17, 18].

PAF is a potential mediator in the coupling of short-lived cell signaling events and long-term genomic responses because it is stimulates immediate early-response gene expression in neuronal and other cells in culture. Furthermore, the use of specific PAF antagonists blocks gene expression in animal models of brain trauma. PAF rapidly and transiently augments, via transcriptional activation, levels of the c-*fos* and c-*jun* transcription factor mRNAs in a neural cell line [34]. Studies involving 5′ deletion mutagenesis of the c-*fos* promoter show that the calcium-response element is necessary for the PAF-induced response. Expression of the mRNAs for c-*fos* and the zinc finger transcription factor *zif*268 is induced by addition of PAF to rat astroglia [7]. PAF also induces early-response gene expression in immune, inflammatory, and other cells, including the heparin-binding epidermal growth factor in monocytes (29), c-*fos* and *egr-2* in lymphoblastoid cell lines [25], the transcription factor NF-[κ] B, and immunoglobulins in human B cell lines [33], c-*fos* and TIS 1 in A-431 epidermoid carcinoma cells [37], and c-*fos* and c-*jun* in rabbit corneal epithelium [2].

A role for PAF in the induction of early-response gene expression in cerebral ischemia or neurotrauma is implied by the inhibitory effects of the intracellular PAF receptor antagonist BN50730. A single electroconvulsive shock rapidly elevates c-*fos* and *zif*268 mRNA levels in the rat hippocampus and cerebral cortex, an effect that is partially blocked by pretreatment of the animals with an intraperitoneal (systemic) or intracerebroventricular (localized) injection of BN50730 in the absence of overt changes in seizure activity [22].

PAF can also induce the inducible form of prostaglandin synthase [PGHS-2, cyclooxygenase-2 (COX-2)]. Prostaglandin synthase catalyzes the rate-limiting step in the oxygenation of arachidonic acid to prostaglandin H_2, the precursor of prostaglandins, and is therefore an important control point for the regulation of lipid second messenger synthesis. The constitutive isoform, PGHS-1, appears to be responsible for prostaglandin synthesis under physiological conditions, while the inducible isoform, PGHS-2, which is encoded by an early-response gene [19, 28], is responsible for the augmentation of prostaglandin synthesis during pathophysiological responses. However, PGHS-2 is expressed constitutively, but is still inducible, in postnatal rat brain [38] and rabbit corneal epithelium [36]. It has recently been shown that PGHS-1 is localized mainly in the endoplasmic reticulum in mouse 3T3 fibroblasts, while PGHS-2 is concentrated in the nuclear envelope [27]. This suggests that PGHS-2 induction may direct prostaglandin action to the nucleus [12]. PGHS-2 is inducible by a number of mitogens, growth factors, and cytokines, and its induction is inhibited by glucocorticoids [14].

PAF can induce PGHS-2 expression in nonneuronal cells and in the brain. When constructs are transfected into cells using the calcium phosphate–calcium coprecipitation procedure in the presence of retinoic acid there is dose-dependent (1–50 n*M*) PAF-induced expression of the reporter, driven by the proximal 371 bp of the mouse PGHS-2 promoter [2]. There is also activity from a 963-bp promoter fragment, although less, suggesting the presence of inhibitory sequences upstream from –371. The effect is rapid, with some activity as early as after 15 or 45 min of incubation with the ligands, suggesting the involvement of preexisting transcription factors. Preincubation of the cells with the intracellular PAF antagonist BN50730 further supports this effect as being a specific receptor-mediated phenomenon. The use of constructs with 5′ deletions of this promoter narrows down the main "PAF-responsive" region of the promoter to between –371 and –300; deletion of this region lowers PAF induction from 31 times to 4.1 times control levels.

In two experimental models of rat brain trauma, seizures, and cryogenically induced vasogenic edema, PGHS-2 mRNA and protein are induced while PGHS-1 expression remains unaltered. Pretreatment of the animals with BN50730 significantly reduces PGHS-2 induction [23]. Unlike the expression of other immediate-early genes in these models, which typically reach peak expression by 1 h after stimulus and return to basal levels within 3–4 h, PGHS-2 expression reaches a peak 6 h after stimulus and remains elevated at 24 h. BN50730 is able to inhibit this induction more effectively than dexamethasone, which is a synthetic glucocorticoid that is a potent inhibitor of PGHS-2 induction in many cells and tissues. This suggests that the PAF antagonist and the glucocorticoid inhibit gene expression via distinct mechanisms.

Conclusion: Potential Significance of Bioactive Lipids to Maturation Phenomenon in Cerebral Ischemia

PAF has multiple effects on excitatory neurotransmission and neural gene expression, and is involved in the modulation of synaptic function in physiological and pathophysiological situations. PLA_2, PAF, and COX-2 induction may represent a signaling pathway transneuronally modulated by excitatory amino acid neurotransmitters. The neuroprotection by presynaptically active PAF receptor antagonists in animal models of cerebral ischemia [11, 30, 31] may be due in part to their inhibitory effects on excitotoxic neuronal damage, as well as to their inhibition of PAF-induced synthesis of inflammatory mediators. In addition, the intracellular PAF antagonist BN50730 inhibits kindling epileptogenesis [26]. Figure 1, depicts PAF involvement in pathways of neuronal plasticity or apoptosis, the induction of cascades of gene expression, and also in inflammatory events, by stimulation of prostaglandin synthesis. Inducible PGHS in ischemic injury and during reperfusion following cerebral ischemia, along with potential effects of prostaglandins as modulators of cerebral circulation, are involved in vasodilation and enhancement of cerebral blood flow [13]. Conversely, the breakdown of the blood-brain barrier and infiltration of inflammatory cells during cerebral edema might allow lipid mediators of extra neural origin to exert their effects on neuronal cells.

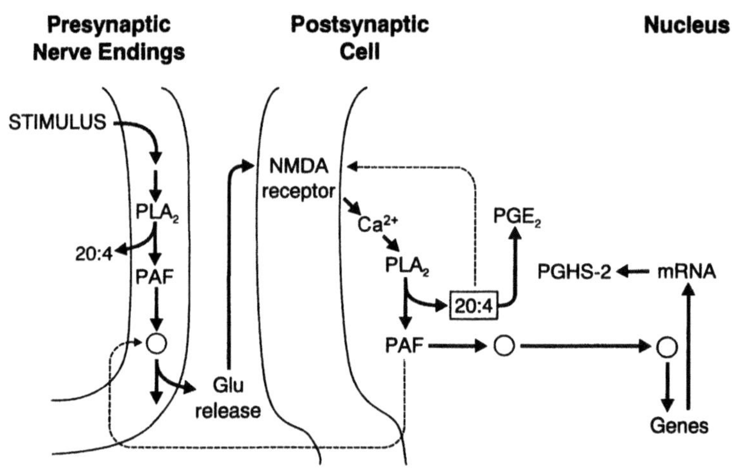

Fig. 1. Transneuronal signaling and bioactive lipids. PAF enhances glutamate release and is involved in long-term potentiation. A depolarizing stimulus at the nerve ending triggers glutamate release. Glutamate occupancy of the postsynaptic NMDA receptor raises intracellular calcium in the postsynaptic neuron. Activation and membrane translocation of the cytoplasmic PLA_2 results in the release of free arachidonic acid and PAF. While these bioactive lipids have very short biological half-lives, upon repeated stimulus sufficient PAF accumulates to diffuse back across the synaptic cleft [16]. PAF binds to its presynaptic receptor and enhances glutamate excytotosis by an as yet undefined mechanism. The free arachidonic acid released by PLA_2 activity can further enhance neurotransmission by acting on the NMDA receptor. During synaptic plasticity events, sufficient PAF might accumulate to activate the gene expression which is probably involved in long-term alterations of synaptic function

Because of the significance of (a) PAF in synaptic plasticity, long-term potentiation, and the formation of certain forms of memory; (b) PAF in excitotoxicity; and (c) COX-2 both in plasticity and in the injury-inflammatory response, we hypothesize that these events play a central role in the enhanced tolerance to ischemic injury. Thus, it is predicted that PLA_2 activation, PAF accumulation, and COX-2 induction should be reduced under these conditions. This hypothesis remains to be tested.

Acknowledgments. This work was supported by funding from the National Institute of Health, grant number 120-NS23002 and the Cooperative Agreement with United States Army Medical Research (DAMD 17-93-V-3013).

References

1. Aveldaño MI, Bazan NG (1975) Rapid production of diacylglycerols enriched in 20:4 and stearate during early brain ischemia. J Neurochem 25: 919–920
2. Bazan HEP, Tao Y, Bazan NG (1993) Platelet-activating factor induces collagenase expression in corneal epithelial cells. Proc Natl Acad Sci USA 90: 8678–8682
3. Bazan NG (1970) Effects of ischemia and electroconvulsive shock on free fatty acid pool in the brain. Biochim Biophys Acta 218: 1–10
4. Bazan NG, Fletcher BS, Herschman HR, Mukherjee PK (1994) Platelet-activating factor and retinoic acid synergistically activate the inducible prostaglandin synthase gene. Proc Natl Acad Sci USA 91: 5252–5256
5. Birkle DL, Bazan NG (1987) Effect of bicuculline-induced status epilepticus on prostaglandins and hydroxyeicosatetraenoic acids in rat brain subcellular fractions. J Neurochem 48: 1768–1778
6. Bonventre JV, Koroshetz WJ (1993) Phospholipase A_2 (PLA_2) activity in gerbil brain: characterization of cytosolic and membrane-associated forms and effects of ischemia and reperfusion on enzymatic activity. J Lipid Mediat 6: 457–471
7. Dell' Albani P, Condorelli DF, Mudo G, Amico C, Bindoni M, Belluardo N (1993) Platelet-activating factor and its methoxy analogue ET-18-OCH3 stimulate immediate-early gene expression in rat astroglial cultures. Neurochem Int: 22: 567–574
8. Doucet JP, Bazan NG (1993) Excitable membranes, lipid messengers and immediated-early genes. Alteration of signal transduction in neuromodulation and neurotrauma. Mol Neurobiol 6: 407–424
9. Francescangeli E, Goracci G (1989) The de novo biosynthesis of platelet-activating factor in rat brain. Biochem Biophys Res Commun 161: 107–112
10. Futaki N, Takahashi S, Yokoyama M, Arai I, Higuchi S, Otomo S (1994) NS-398, a new anti-inflammatory agent, selectively inhibits prostaglandin G/H synthase/cyclooxygenase (COX-2) activity in vitro. Prostaglandins 47: 55–59
11. Gilboe DD, Kintner D, Fitzpatrick JH, Emoto SE, Esanu A, Braquet PG, Bazan NG (1991) Recovery of postischemic brain metabolism and function following treatment with a free radical scavenger and platelet activating factor. J Neurochem 56: 311–319
12. Goetzl EJ, An S, Smith WL (1995) Specificity of expression of eicosanoid mediators in normal physiology and human disease. FASEB J 9: 1051–1058
13. Hallenbeck JM, Furlow TW (1979) Prostaglandin I_2 and indomethacin prevent inhibition of post-ischemic brain reperfusion in the dog. Stroke 10: 629–637
14. Herschman HR (1994) Regulation of prostaglandin synthase-1 and prostaglandin synthase-2. Cancer Metastasis Rev 13: 241–256
15. Honda Z, Nakamura M, Miki I, Minami M, Wantabe T, Seyama Y, Okado H, Toh H, Ito K, Miyamoto T, Shimizu T (1991) Cloning by functional expression of platelet-activating factor receptor from guinea-pig lung. Nature 394: 342–346
16. Kato K, Clark GD, Bazan NG, Zormuski CF (1994) Platelet-activating factor as a potential retrograde messenger in CA1 hippocampal long-term potentiation. Nature 367: 179–182

17. Kiessling M, Gass P (1993) Immediate early gene expression in experimental epilepsy. Brain Pathol 3: 381–393
18. Kogure K, Kato H (1993) Altered gene expression in cerebral ischemia. Stroke 24: 2121–2127
19. Kujubu DA, Fletcher BS, Varnum BC, Lim RW, Herschman HR (1991) TIS10, a phorbol ester tumor promoter-inducible mRNA from Swiss 3T3 cells, encodes a novel prostaglandin synthase/ cyclooxygenase homologue. J Biol Chem 266: 12866–12872
20. Kumar R, Harvey S, Kester N, Hanahan D, Olson M (1988) Production and effects of platelet-activating factor in the rat brain. Biochim Biophys Acta 963: 375–383
21. Kunz D, Gerard NP, Gerard C (1992) The human leukocyte PAF receptor, cDNA cloning, cell surface expression and construction of a novel epitope-bearing analog. J Biol Chem 267: 9101–9106
22. Marcheselli VL, Bazan NG (1994) Platelet-activating factor is a messenger in the electroconvulsive shock-induced transcriptional activation of c-*fos* and *zif*-268 in hippocampus. J Neurosci Res 37: 54–61
23. Marcheselli VL, Bazan NG (1996) Sustained induction of prostaglandin endoperoxide synthase-2 by seizures in hippocampus: inhibition by a platelet-activating factor antagonist. J Biol Chem 271: 24794–24799
24. Marcheselli VL, Rossowska M, Domingo MT, Braquet P, Bazan NG (1990) Distinct platelet-activating factor binding sites in synaptic endings and in intracellular membranes of rat cerebral cortex. J Biol Chem 265: 9140–9145
25. Mazer B, Domenico J, Sawami H, Gelfand EW (1991) Platelet-activating factor induces an increase in intracellular calcium and expression of regularity genes in human B lymphoblastoma cells. J Immunol 146: 1914–1920
26. Moises J, Allan G, Marcheselli VL, Mahandere V, Bazan NG (1994) A PAF antagonist inhibits the expression of hippocampal immediate early genes in a kindling model of epileptogenesis. Epilepsia 35: 139
27. Morita I, Schindler M, Regier MK, Otto JC, Hori T, DeWitt DL, Smith WL (1995) Different intracellular locations for prostaglandin endoperoxide H synthase-1 and -2. J Biol Chem 270: 10902–10908
28. O'Banion MK, Sadowski HB, Winn V, Young DA (1991) A serum- and glucocorticoid-regulated 4-kilobase mRNA encodes a cyclooxygenase-related protein. J Biol Chem 266: 23261–23267
29. Pan Z, Kravchenko VV, Ye RD (1995) Platelet activating factor stimulates transcription of the heparin-binding epidermal growth factor-like growth factor in monocytes. Correlation with an increased kappa B binding activity. J Biol Chem 270: 7787–7790
30. Panetta T, Marcheselli VL, Braquet P, Spinnewyn B, Bazan NG (1987) Effects of a platelet-activating factor antagonist (BN 52021) on free fatty acids, diacylglycerols, polyphosphoinositides and blood flow in the gerbil brain: inhibition of ischemia-reperfusion induced cerebral injury. Biochem Biophys Res Commun 149: 580–587
31. Prehn JHM, Krieglstein J (1993) Platelet-activating factor antagonists reduce excitotoxic damage in cultured neurons from embryonic chick telencephalon and protect the rat hippocampus and neocortex from ischemic injury in vivo. J Neurosci Res 34: 179–188
32. Prescott SM, Zimmerman GA, McIntyre TM (1990) Platelet-activating factor. J Biol Chem 265: 17381–17384
33. Smith CS, Shearer WT (1994) Activation of NF-kappa B and immunoglobulin expression in response to platelet-activating factor in a human B cell line. Cell Immunol 155: 292–303
34. Squinto SP, Block AL, Braquet P, Bazan NG (1989) Platelet-activating factor stimulates a Fos/Jun/ AP-1 transcriptional signaling system in human neuroblastoma cells J Neurosci Res 24: 558–566
35. Sugimoto T, Tsuchimochi H, McGregor CG, Mutoh H, Shimizu T, Kurachi Y (1992) Molecular cloning and characterization of the platelet-activating factor receptor from human heart. Biochem Biophys Res Commun 189: 617–624
36. Tao Y, Bazan NG, Bazan HEP (1995) Platelet-activating factor (PAF) increases prostaglandin synthase-2 (COX-2) gene expression in corneal epithelium. Arvo Abstr 36: S573
37. Tripathi Y, Kandala J, Guntaka R, Lim R, Shukla S (1991) Platelet-activating factor induces expression of early response genes c-*fos* and TIS-1 in human epidermoid carcinoma-432 cells. Life Sci 49: 1761–1767

38. Yamagata K, Andreasson KI, Kaufmann WE, Barnes CA, Worley PF (1993) Expression of a mitogen-inducible cyclooxygenase in brain neurons: Regulation by synaptic activity and glucocorticoids. Neuron 11: 371–386
39. Yoshijara Y, Watanabe Y (1990) Translocation of phospholipase A_2 from cytosol to membranes in rat brain induced by calcium ions. Biochem Biophys Res Commun 170: 484–490

Ischemic Neuronal Injury and Gene Expression of Facilitative and Inhibitory Growth Factors

K. Yamada[1], A. Masago[1], A. Iwata[1], E. Kohmura[2], T. Yuguchi[2], T. Sakaki[2], and T. Hayakawa[2]

Summary

We studied the regulatory mechanism of gene expression for the facilitative growth factor, basic fibroblast growth factor (bFGF), and the growth inhibitory factor (GIF) in ischemic brain tissue. The mRNA expression of these growth factors was studied using the in situ hybridization method and compared with the expression of c-*fos*, c-*jun*, and heat shock protein(HSP) 70 mRNA in the serial sections of the brain tissue during the recirculation period after 2 h-occlusion of the middle cerebral artery in rat. The pattern of gene expression for bFGF, c-*fos*, and c-*jun* is quite similar, whereas HSP70 and GIF gene expression is somewhat different from others. The result of these findings suggests that gene expression of facilitative and inhibitory growth factors is well controlled in order. In the same tissue preparation, we studied gene expression of neuropsin, a hippocampus-specific serine protease. Down-regulation of neuropsin after ischemic insult was found and it may be related to the restorative process of hippocampal neurons against ischemic stress.

Introduction

The aim of this study is to analyze the regulatory mechanism of gene expression for the facilitative growth factor, basic fibroblast growth factor (bFGF), and the growth inhibitory factor (GIF) in ischemic brain tissue. The mRNA expression of these growth factors was compared to the expression of c-*fos*, c-*jun*, and heat shock protein (HSP) 70 mRNA in the serial sections of the brain tissue during the recirculation period following temporary occlusion of the middle cerebral artery.

In the same tissue preparation, we studied gene expression of neuropsin, a hippocampus-specific serine protease [2], in relation to c-*fos*, c-*jun* and HSP70 mRNA expression. With those studies we obtained information for a thorough understanding of the regenerative process of injured neurons.

[1] Department of Neurosurgery, Nagoya City University Medical School, 1 Kawasumi, Mizuho-ku, Nagoya 467, Japan
[2] Department of Neurosurgery, Osaka University Medical School, 2 Yamadaoka, Suita 565, Osaka, Japan

Maturation Phenomenon in Cerebral Ischemia II
U. Ito et al. (Eds.)

Materials and Methods

Animal Preparation

Using Wistar rats, the left middle cerebral artery of was occluded for 2 h by the thread occlusion method and reperfused for 1 h, 3 h, 6 h, 12 h, 1 day, 2 days, and 5 days. The brains were then frozen, thin-sliced and prepared for in situ hybridization of bFGF, c-*fos*, c-*jun* and HSP70. In the alternative experiment set-up, the left middle cerebral artery of the Wistar rats was occluded using Tamura's method and brains were removed at 6 h, 1 day, 3 days, 4 days, 7 days, and 14 days after occlusion. They were used for in situ hybridization for GIF mRNA.

In Situ Hybridization

We used oligo-DNA probes for in situ hybridization of bFGF c-*fos*, c-*jun* and HSP70. The specification of the probes is as follows: the rat bFGF ologoprobe (30 mer) complementary to the putative exon 2 (obtained from Oncogene Research); HSP70 ologoprobe (30-mer) complementary to the base number 539–568 of the rat HSP70 cDNA [7]; c-*fos* oligoprobe (51-mer) complementary to amino acid 60–67 of the rat c-*fos* protein [3]; and c-*jun* probe (45-mer) complementary to the last 15 amino acid residues of the predicted c-*jun* protein [1].

The GIF and neuropsin mRNA expression was studied with RNA probes transcribed by their cDNA. We used the unique 3′-non-coding sequence of the rat GIF cDNA which was cloned by Kobayashi et al. [4]. Neuropsin cDNA was supplied by Shiosaka [2] and used for this study.

The techniques of in situ hybridization for oligo-DNA probes and RNA probes were identical to those of our previous report [6, 8, 11, 12].

Results

Cortical and Subcortical Gene Expression

Expression of bFGF mRNA began to be upregulated as early as 6 h after reperfusion and lasted 2 days (Fig. 1). Upregulation of bFGF mRNA expression is found in the cortex of the ischemic hemisphere and peri-infarcted white matter (Fig. 1C,D). The pattern of bFGF mRNA expression is similar to those of c-*fos* and c-*jun*, suggesting that the same regulatory mechanism, such as spreading depression, might take place (Fig. 2). The HSP70 mRNA expression is localized to the tissue exposed to ischemia and is different from that of bFGF, c-*fos*, and c-*jun*, suggesting that another mechanism may take place for gene expression of HSP70.

Hippocampal Gene Expression

Gene expression of bFGF, c-*jun*, c-*fos* and HSP70 was detectable in the hippocampal pyramidal layer and dentate gyrus (Fig. 2). The bFGF mRNA was constitutively ex-

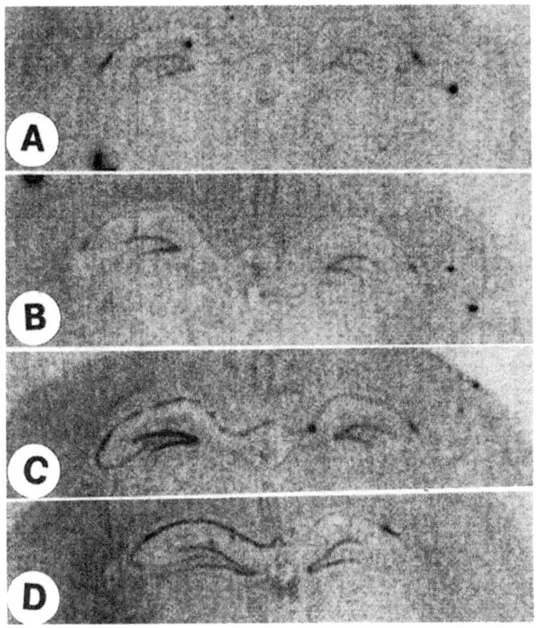

Fig. 1A–D. Expression of fibroblast growth factor (FGF) mRNA in the thread occlusion model of the rat middle cerebral artery. The *left-hand side* is the side of ischemia. **A**, sham control; **B**, 2-h occlusion and 3-h reperfusion; **C**, 2-h occlusion and 6-h reperfusion; **D**, 2-h occlusion and 12-h reperfusion

pressed in the dentate gyrus of the hippocampus but only slightly expressed in the pyramidal cell layer (Fig. 1). When 2h of thread occlusion took place and reperfusion occurred, FGF mRNA expression was upregulated in the pyramidal cell layer of the lesion-side hippocampus as early as 3h after reperfusion, and the expression extended to the contralateral pyramidal cell layer (Fig. 1).

Gene Expression of Growth Inhibitory Factor

The gene expression of the GIF is quite different from other growth factors or immediate early genes. Expression of the GIF mRNA was suppressed in the lesion-side cortex at 1 day after clip occlusion of the middle cerebral artery (Fig. 3A). The GIF mRNA expression was upregulated transiently in the lesion-side cortex at 4 days after ischemia (Fig. 3B), but it returned to the baseline level 1 week after ischemia. The GIF mRNA expression was increased in the narrow area of periischemic gliotic tissue 2 weeks after ischemia.

Gene Expression of Neuropsin

Expression of neuropsin mRNA was different from those of immediate early genes and HSP70. The neuropsin gene expression in the hippocampus was a mirror image

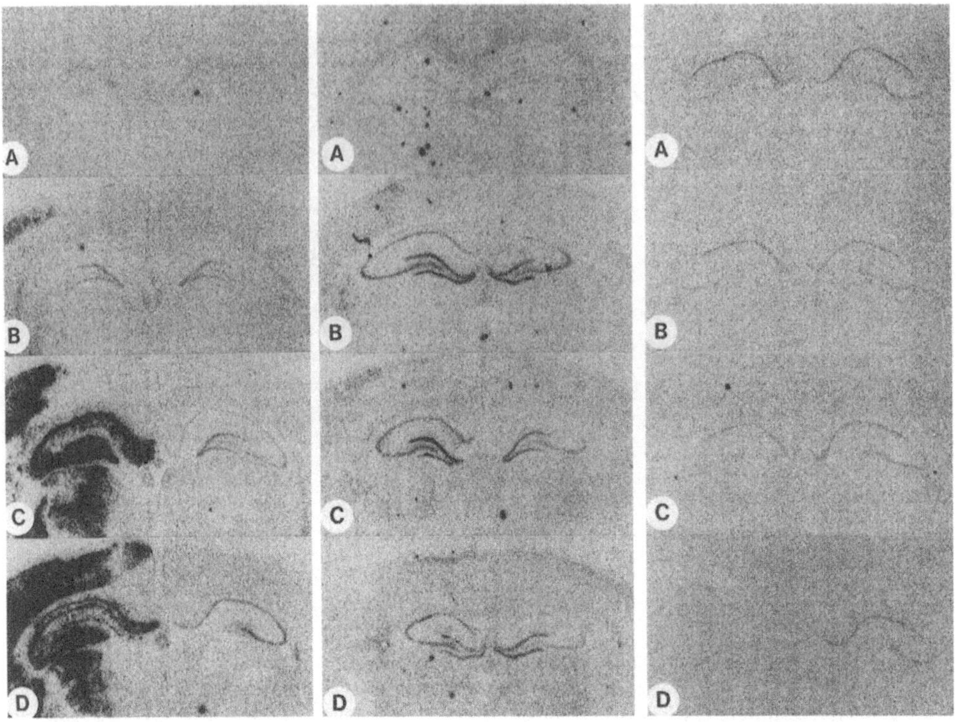

Fig. 2A–D. Messenger RNA expression of heat shock protein (HSP) 70, *c-jun* and neuropsin in the thread occlusion model of the middle cerebral artery. The *left hand side* is the side of ischemia. **A**, sham control; **B**, 2-h occlusion and 1-h reperfusion; **C**, 2-h occlusion and 3-h reperfusion; **D**, 2-h occlusion and 6-h reperfusion

of the HSP70 and *c-jun* gene expression (Fig. 2). The neuropsin gene expression decreased 3h after ischemia and was completely suppressed in the pyramidal cell layer of the lesion-side hippocampus at 6–48h after reperfusion (Fig. 2). It returned to the normal level in 72h after reperfusion when other upregulated genes returned to the normal level.

Discussion

The present report clearly indicated that one of the facilitative growth factors, bFGF is indeed upregulated in the periischemic cortex and white matter. We have previously reported that intracisternal injection of bFGF is effective for prevention of thalamic degeneration after cortical infarction [9] or cortical ablation [5]. The question arises why endogenous bFGF can not prevent thalamic degeneration if bFGF mRNA is upregulated in the peri-injury tissue. Our previous study indicated that immunoreactivity of bFGF started to be detectable 4 days after ischemia [10]. Therefore, a "time lag" between mRNA expression and protein detection is present, suggesting a rationale of exogenous bFGF administration for the treatment of ischemic neuronal injury.

The pattern of bFGF mRNA expression is quite similar to that of c-*fos* and c-*jun* but somewhat different from HSP70 expression (Figs. 1, 2). Furthermore, the expression of bFGF, c-*fos* and c-*jun* was suppressed by MK801, but HSP70 was not (unpublished observation). The result indicated that expression of bFGF as well al c-*fos* and c-*jun* is regulated by spreading depression.

The GIF mRNA expression was downregulated 1 day after ischemia but transiently upregulated 4 days after ischemia. This pattern of GIF expression may correlate well with the activity of axonal regeneration. One day after ischemia, neurons adjacent to the ischemic core may have neurites injured by ischemia, and neurite outgrowth might be necessary, resulting in down regulation of GIF mRNA. Four days after ischemia, primary response to neurite injury might be completed and GIF mRNA might be upregulated to stop neurite outgrowth. These data suggest a role of GIF in the regulation of neurite outgrowth.

The hippocampal gene expression in the thread occlusion model is controversial. In this model the posterior communicating artery and the posterior cerebral artery which supply blood to the hippocampal formation might be compromised. The HSP70 mRNA expression was strongly upregulated in the lesion-side hippocampus, indicating ischemic compromise to the hippocampal formation. Neuronal death, however, was not definitive when we examined tissue sections 5 days after 2-h temporary thread occlusion (unpublished observation). Therefore, ischemia to the hippocampus exerts sublethal stress on the neurons. Indirect ischemic effect through the limbic system is also possible for hippocampal gene expression.

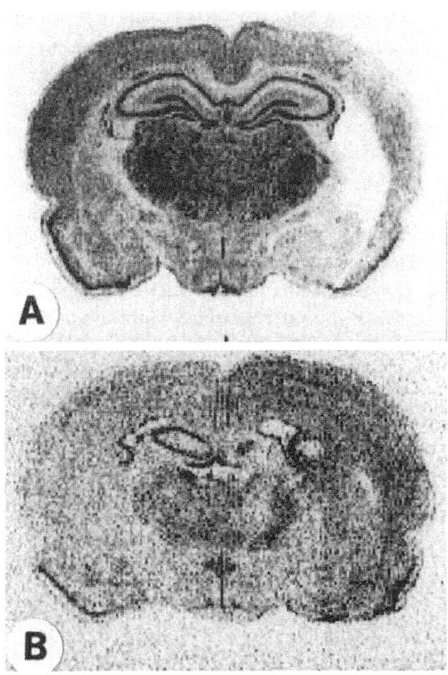

Fig. 3A,B. Messenger RNA expression of the growth inhibitory factor (GIF) in the clip occlusion model of the rat middle cerebral artery. The *right-hand side* is the side of ischemia. **A**, 1 day after ischemia; **B**, 4 days after ischemia

No matter what the mechanism of hippocampal gene expression might be, it is interesting to find mRNA downregulation of neuropsin, a hippocampus specific serine protease. The neuropsin is a family of serine protease such as tissue-type plasminogen activator (tPA) or urine-type plasminogen activator (uPA). Different from tPA or uPA, neuropsin is only expressed in the hippocampus. The exact mechanism of downregulation in the postischemic period remains unclear. Chen et al. [2] reported that electrical stimulation to the hippocampus causes upregulation of the neuropsin mRNA expression, whereas low frequency electrical stimulation to the hippocampus downregulates neuropsin mRNA expression. Further study is needed for clarification of this matter.

References

1. Angel P, Allegretto EA, Okino ST, Hattori K, Boyle WJ, Hunter T, Karin M (1988) Oncogene jun encodes a sequence-specific transactivator similar to AP-1. Nature 322: 166–171
2. Chen ZL, Yoshida S, Kato K, Momota Y, Suzuki J, Tanaka T, Ito J, Nishino H, Aimoto S, Kiyama H et al (1995) Expression and activity-dependent changes of a novel limbic-serine protease gene in the hippocampus. J Neurosci 15: 5088–5097
3. Curran T, Gordon MB, Rubino KL, Sambucetti LC (1987) Isolation and characterization of the c-fos (rat) cDNA and analysis of post-translational modification in vitro. Oncogene 2: 79–84
4. Kobayashi H, Uchida Y, Ihara Y, Nahajima K, Kohsaka S, Miyatake T, Tsuji S (1993) Molecular cloning of rat growth inhibitory factor cDNA and the expression in the central nervous system. Mol Brain Res 19: 188–194
5. Kohmura E, Yuguchi T, Yamada K, Sakaguchi T, Hayakawa T (1994) Recombinant basic fibroblast growth factor spares thalamic neurons from retrograde degeneration after ablation of the somatosensory cortex in rats. Restr Neurol Neurosci 6: 309–316
6. Kohmura E, Yuguchi T, Yamada K, Sakaguchi T, Wanaka A, Hayakawa T (1995) Expression of c-fos mRNA after cortical ablation in rat brain is modulated by basic fibroblast growth factor (bFGF) and the NMDA receptor is involved in c-fos expression. Mol Brain Res 28: 117–121
7. Longo FM, Wang S, Narasimhan P, Zhang JS, Chen J, Massa SM, Sharp FR (1993) cDNA cloning and expression of stress-inducible rat hsp70 in normal and injured rat brain. J Neurosci Res 36: 325–335
8. Sakaguchi T, Yamada K, Wanaka A, Kohmura E, Yuguchi T, Taneda M, Kataoka K, Tohyama M, Hayakawa T (1994) Expression of basic fibroblast growth factor receptor messenger RNA in the periinfarcted brain tissue. Restr Neurol Neurosci 7: 29–36
9. Yamada K, Kinoshita A, Kohmura E, Sakaguchi T, Taguchi J, Kataoka K, Hayakawa T (1991) Basic fibroblast growth factor prevents thalamic degeneration after cortical infarction. J Cereb Blood Flow Metabol 11: 472–478
10. Yamada K, Kohmura E, Taguchi J, Kinoshita A, Sakaguchi T, Hayakawa T (1992) Expression of basic FGF in periinfarcted brain tissue and modification of postischemic thalamic degeneration by exogenous basic FGF. In: Globus MYT, Dietrich WD (eds) The role of neurotransmitters in brain injury. Plenum, New York, pp 259–264
11. Yuguchi T, Kohmura E, Yamada K, Wanaka A, Otsuki H, Sakaguchi T, Yamashita T, Tohyama M, Hayakawa T (1994) Messenger RNA and protein expression of basic fibroblast growth factor receptor after cortical ablation. Mol Brain Res 25: 50–56
12. Yuguchi T, Kohmura E, Yamada K, Sakaki T, Yamashita T, Otsuki H, Wanaka A, Tohyama M, Tsuji S, Hayakawa T (1995) Changes in growth inhibitory factor mRNA expression compared with those in c-jun mRNA expression following facial nerve transection. Mol Brain Res 28: 181–185

Is Apoptosis Involved in Glutamate Neurotoxicity?

J. Ikeda[1,2], S. Terakawa[3], K. Asai[2], I. Morita[1], S. Murota[1], and K. Hirakawa[4]

Summary

Recent studies on glutamate neurotoxicity suggest that apoptosis plays a role in excitotoxic neuronal death. The present study was designed to examine the possible involvement of apoptosis in such death.

In the first part of the study we employed Nomarski optics combined with a digital image processor to examine the acute type of glutamate neurotoxicity, and found a rapid change in the nucleus followed by cellular swelling. These changes could be prevented by treatment with MK-801, a *N*-methyl-D-aspartate (NMDA) receptor antagonist. These results indicate that morphological changes mediated by overactivation of NMDA receptor does not fulfill the morphological criteria of apoptosis, but do suggest that the nuclear disintegration is involved as a leading step in acute-type glutamate excitotoxicity.

In the second part of the study, we examined the effects of a peptide inhibitor for interleukin-1β-converting enzyme (ICE) and a free radical scavenger (MCI-186) on the delayed type of glutamate neurotoxicity. Whereas the ICE inhibitor failed to show a beneficial effect, MCI-186 significantly reduced the excitotoxic damage, suggesting that apoptosis may be involved in the delayed type of neuronal death, and that free radical production may be related to this process.

Introduction

Excitotoxic neuronal death, which is mediated by overactivation of the glutamate receptor, has been implicated in several neurological disorders such as stroke, epilepsy, and head injury [6, 8, 22]. Among the three types of glutamate receptors, activation of the *N*-methyl-D-asparate (NMDA) subtype and the subsequent calcium influx has been attributed to neuronal injury [3]. However, the downstream events culminating in irreversible injury remain unclear. Recently, characteristics of

[1] Department of Physiological Chemistry, Graduate School, Tokyo Medical and Dental University, 1-5-45 Yushima Bunkyo-ku, Tokyo 113, Japan
[2] Asai Hospital, 38-1 Katoku, Togane-shi Chiba 283, Japan
[3] Photon Medical Center Hamanatsu University School of Medicine, 3600 Handa-maclin Shizuoka 431-31, Japan
[4] Department of Neurosurgery, Tokyo Medical and Dental University, 1-5-45 Yushima Bunkyo-ku Tokyo 113, Japan

Maturation Phenomenon in Cerebral Ischemia II
U. Ito et al. (Eds.)

apoptosis such as internucleosomal DNA fragmentation and the protective effect of protein synthesis inhibitors have been found in excitotoxic neuronal death, both in vitro or in vivo [16, 24]. Also, a rise in intracellular calcium concentration was shown to initiate the apoptotic process [17]. Therefore, apoptosis, which was previously though to be involved in physiological neuronal death during development, may play an important role in excitotoxic neuronal death.

The two major types of cell death, namely, apoptosis and necrosis, are distinguished on the basis of morphological abnormalities of the cells. Apoptosis is characterized by cell shrinkage and nuclear condensation with rather nonspecific ultrastructural alterations of the cytoplasm. The nucleus undergoes a relatively characteristic change, namely, chromatin condensation and subsquent DNA fragmentation [5]. On the other hand, an earlier study on excitotoxic neuronal death by phase-contrast microscopy showed acute swelling, darkening, and an enhanced granulation of the soma, which are consistent with a necrotic type of death [2]. These differences between nuclear changes in apoptotic cell death and cytoplasmic changes in excitotoxic cell death prompted us to define more precisely the glutamate-induced neuronal changes, especially in the nucleus. As the first part of our present study, we employed a video-enhanced differential interference contrast (DIC, Nomarski) microscope to examine the morphological changes at a high resolution, and found that glutamate induced a dramatic change in the nucleus as the earliest response preceding any morphological changes in the cytoplasm.

Despite its importance in normal physiology, apoptosis is poorly understood in terms of the molecular and biochemical mechanism involved. Recent knowledge on the genetic regulation of cell death was obtained from a study using the nematode *Caenorhabditis elegans* as a model organism [7]. In this worm, three major genes have been identified. Two of them, ced-3 and ced-4, must function in dying cells for death to occur, and another gene, ced-9, is required for prevention of cell death. In mammals, ced-3 and ced-9 were found to exhibit significant homology to the interleukin-1β-converting enzyme (ICE) and bcl-2, respectively [9, 28]. The mechanism of action of these genes is not clear. However, several lines of evidence suggest that products of the bcl-2 gene intervene at some step in the cell death pathway by neutralizing reactive oxygen species [10, 14]. Therefore, as the second part of our study, we investigated the possible involvement of apoptosis-related genes in excitotoxic neuronal death, by testing the effect of an ICE inhibitor (ZVAD-FMK, Enzyme System Products, Livermore, USA) and a free radical scavenger (MCI-186, Mitsubishi Kasei Corporation, Yokohama, Japan) on glutamate neurotoxicity.

Materials and Methods

Cell Culture

Cortical neuron cultures were prepared as previously described [11]. The cortices of fetal rats were dissected and incubated with trypsin (0.25%) and DNase (0.02%) at 37°C for 15 min. The cells were seeded onto polyethylenimine-coated cover slips and cultured for 12 days in Dulbeccos's modified Eagle's medium (DMEM) supplemented with 10% fetal bovine serum.

Morphological Observations

Observations of cultured neurons were performed as described [13]. In brief, a coverslip bearing neurons was placed on the stage of an inverted DIC microscope (Axiovert 35, Zeiss, Oberkochen). The chamber was continuously perfused with calcium (0.9 m*M*) containing phosphate buffer (PBS) (+). The microscope was equipped with a ×100 DIC objective lens and a ×2.5 insertion lens and the DIC image of the cells was obtained with a charge-coupled device (CCD) camera (SSC-M350, Sony, Tokyo). The video image was contrast-enhanced digitally in real time, monitored, and recorded on videotape. After control images had been recorded, the cells were perfused with PBS (+) containing 1 m*M* glutamate. In some experiments, the effect of glutamate was removed after 5 min by perfusing the chamber with glutamate-free PBS (+).

Assessment of Cellular Injury

Neuronal cell injury was assessed 24 h after 5-min exposure of the cells to 200 μM glutamate [12]. Cortical cells grown on over slips were placed on an interactive laser cytometer (ACAS ULTIMA, Meridian Instruments, Inc., Okemos, MI, USA). Surviving neurons were stained with fluorescent diacetate (FDA, 2 mg/ml) and dead ones were stained with propidium iodide (PI, 2 mg/ml), Finally, the ratio of surviving neurons to dead ones was calculated with the microcomputer of the ACAS. In experiments with ICE inhibitor (5 μM) and MCI-186 (5 μM), the drugs were present 30 min before, during, and after exposure to glutamate.

Results and Discussion

Neurons in the control samples had a large nucleus with an amorphous appearance (Fig. 1A). The cytoplasm was granular with small particles exhibiting an occasional movement that indicated transport activity. An apparent change in morphology upon 1-m*M* glutamate exposure was first encountered in the nucleus, which consisted of an increase in granularity followed by swelling (Fig. 1B). Although the time of onset of these change varied slightly from culture to culture, the nuclear change was always observed before the cytoplasmic swelling, which occurred just after the nuclear swelling (Fig. 1C). Finally, small particles appeared in the nucleus and moved everywhere between the granules, suggesting nuclear lysis (Fig. 1D). At this time, the morphological change was irreversible. By 6 h, almost all the neurons had become round with a liquefied nucleus and were positively stained with propidium iodide, indicating that these cells were dead (data not shown).

When the duration of glutamate application was limited to 5 min, the morphological change was reversible (Fig. 2A–F). Thus, despite the formation of granules and subsequent swelling of the nucleus and cytoplasm, cells returned to their initial appearance 1 h after removal of the glutamate. A subsequent challenge with glutamate again induced the above morphological changes (Fig. 2F). All of these changes were blocked by the treatment of the neurons with MK-801 (10 μM).

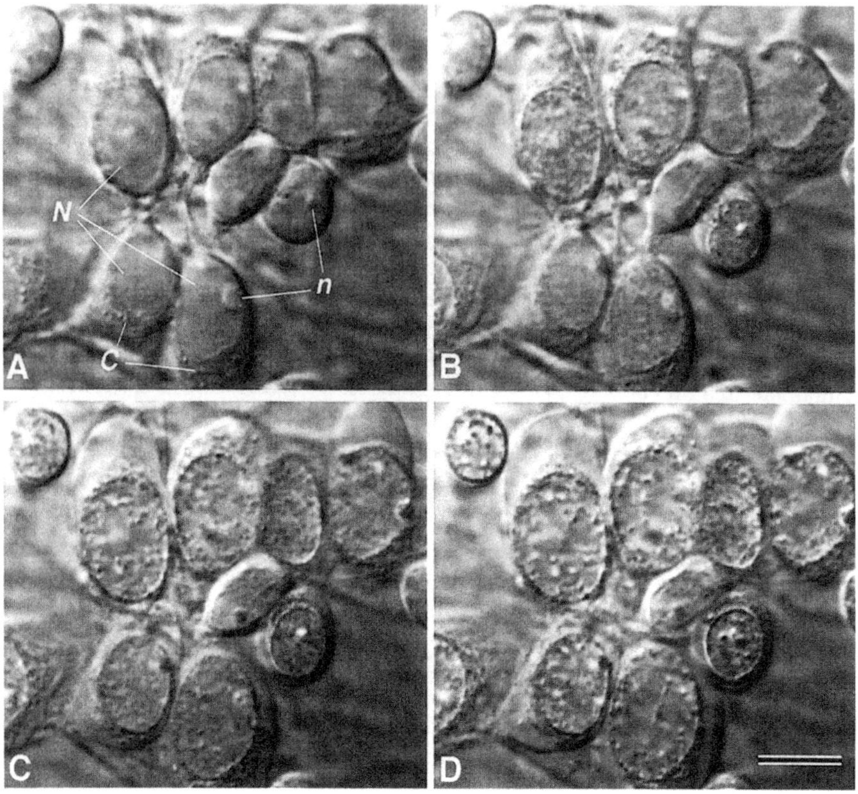

Fig. 1A–D. Effect of 1 m*M* glutamate on morphological properties of neurons observed under a video enhanced contrast-differential interference contrast (VEC-DIC) microscope. Images were taken immediately before (**A**), 3 min (**B**), 10 min (**C**), and 20 min (**D**) after glutamate administration. *N*, nucleus; *n*, nucleoli; *C*, cytoplasm. *Bar* = 10 μM

Table 1. Effects of interleukin-1β-converting enzyme (ICE) inhibitor and MCI-186 on the viability (mean ± SEM) of glutamate-treated cortical neurons

	Control	ICE inhibitor (ZVAD-FMK)	MCI-186
% Survival	54.1 ± 2.2	48.2 ± 2.6	81.2 ± 1.8

Neuronal injury was assessed 24 h after a 5-min exposure of cortical neurons to 200 μM glutamate (see "Materials and Methods" for further details).

Exposure to 200 μM glutamate for 5 min resulted in a reduction in cell survivability to 55% within 24 h. Addition of MCI-186 at concentrations of 5–100 μM to sister cultures prior to the insult substantially ameliorated the neuronal injury, with maximal protection seen at a 50-μM concentration of the drug (82% survival rate). ICE inhibitor failed to protect neurons from glutamate neurotoxicity (Table 1).

In the present study, we used two different concentrations of glutamate because Choi previously recognized two types of excitotoxic neuronal death (i.e., acute and delayed) on the basis of a difference in time course and in ionic dependence [2]. In the first part of our study, morphological changes in acute neuronal death induced by continuous application of 1 m*M* glutamate was investigated. In this excitotoxic paradigm, we found that swelling of the nucleus, followed by cytoplasmic swelling, was the

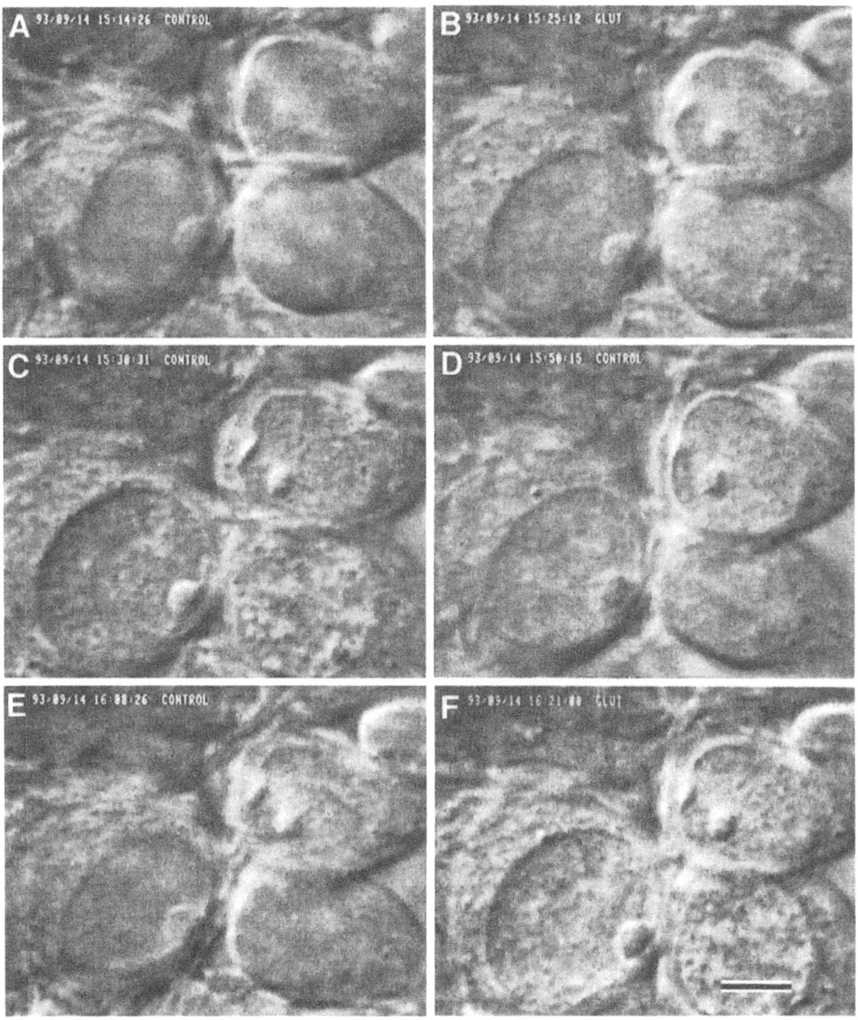

Fig. 2A–F. Reversible effect of glutamate on nuclei of neurons. Video enhanced contrast-differential interference contrast (VEC-DIC) images were taken before (**A**) and 5 min after (**B**) administration of 1 m*M* glutamate. Glutamate was removed immediately after the image in **B**, and images were taken consecutively 5 min (**C**), 25 min (**D**), and 43 min (**E**) thereafter. Then, 1 m*M* glutamate was administered again and an image (**F**) was taken 10 min later. *Bar* = 5 μM

major change in the pathway to acute neuronal death, indicating that apoptosis is not involved in this process. The protective effect of MK-801 indicated that the observed neuronal death was mediated by overstimulation of the NMDA receptors. These results do not appear compatible with those of a previous study, in which internucleosomal DNA fragmentation was observed after continuous exposure to 1 m*M* glutamate in cortical neurons cultured for 4 days [16]. This apparent discrepancy may reflect a difference in maturity of the neurons because NMDA receptors do not play much of a role in glutamate-induced cell death in younger cultures [21].

According to Choi, cytosolic swelling in an acute injury has been attributed to a sodium influx through the plasma membrane [4]. However, in our study, swelling of the cytoplasm was always observed after the irreversible swelling of the nucleus. We speculate that degradation of intranuclear molecules may supply osmotic equivalents to the cytoplasm and contribute in concert with the sodium influx to the cytoplasmic swelling. In support of this view, generation of idiogenic osmotically active particles in brain tissue has been suggested to play a role in ischemic brain edema [23]. Taken together, the morphological changes observed do not fulfill the criteria of apoptosis, but the nuclear disintegration appears to be involved as a leading step in the acute type of excitotoxic neuronal death.

In the second part of our study, we assessed the involvement of apoptosis-related genes in the excitotoxic neuronal death of the delayed type. Application of 100–500 μM glutamate for 5 min is known to induce neuronal death within 12–24 h [20]. It has been well established that the excitotoxic neuronal death in this paradigm is dependent of Ca^{2+} influx through NMDA receptors [3], but the pathomechanism of downstream events remains to be solved. In this regard, a recent finding suggesting the implication of apoptosis-related genes in neuronal death is attractive [25].

ICE is a protease which converts the 33-kDa protease form of IL-1β to the active 17.5-kDa form. There are several pieces of evidence indicating a role of proteases in apoptosis in the central nervous system [15]. For example, ICE inhibitors rescue vertebrate motoneuronal death as a result of trophic factor deprivation [19]. In the present study, we used a peptide inhibitor that mimics ICE substrates and thus potentially inhibits ICE proteolytic activity [18]. We failed to observe any protective effect of the drug on excitotoxic neuronal death, suggesting that ICE is not involved in the pathomechanism whereby Ca^{2+} overinflux kills neurons, although this might be due to the potentially short half-life of the peptide inhibitor.

To probe a role of anti-apoptotic proto-oncogene bcl-2, we used a newly synthesized free radical scavenger MCI-186. Overexpression of bcl-2 has been reported to protect neuronal death induced by glutamate [1]. The protective effect exhibited by bcl-2 has been attributed to its ability to function as an antioxidant [10, 14]. We found that MCI-186, which has been shown to scavenge hydroxy radicals [26], significantly reduce excitotoxic neuronal death. Together with the recent work suggesting that generation of free radicals may be involved in apoptosis [27], it is possible that MCI-186 protects the neurons from apoptosis by interfering with the oxidant pathway.

In summary, the present study was designed to provide a clue for the cause of apoptosis in glutamate neurotoxicity. Although the morphological changes observed do not fulfill the criteria of apoptosis in the acute type of neuronal death, we found a possible involvement of apoptosis in the delayed type of neuronal death, which process may be related to free radical production.

References

1. Behl C, Hovey L, Krajewski S, Schubert D, Reed JC (1993) Bcl-2 prevents killing of neuronal cells by glutamate but not by β protein. Biochem Biophys Res Commun 197: 949–956
2. Choi DW (1987) Ionic dependence of glutamate neurotoxicty. J Neurosci 7: 369–379
3. Choi DW, Koh J-Y, Peters P (1988) Pharmacology of glutamate neurotoxicity in cortical cell culture: attenuation by NMDA antagonists. J Neurosci 8: 185–196
4. Choi DW (1988) Glutamate neurotoxicity and diseases of the nervous system. Neuron 1: 623–634
5. Clarke PGH (1990) Developmental cell death: morphological diversity and multiple mechanisms. Anat Embryol (Berl) 181: 195–213
6. Dichter MA (1989) Cellular mechanisms of epilepsy and potential new treatment strategies. Epilepsia 30: S3-S12
7. Ellis RE, Yuan J, Horvitz HR (1991) Mechanisms and functions of death. Annu Rev Cell Biol 7: 663–698
8. Faden AI, Demediuk P, Panter S, Vink P (1989) The role of excitatory amino acids and NMDA receptors in traumatic brain injury. Science 244: 798–800
9. Hengartner MO, Horvits HR (1994) C. elegans cell survival gene ced-9 encodes a functional homologue of the mammalian proto-oncogene bcl-2. Cell 76: 665–676
10. Hockenbery DM, Oltvai ZN, Yin X-M Milliman CL, Korsmeyer SJ (1993) Bcl-2 functions in an antioxidant pathway to prevent apoptosis. Cell 75: 241–251
11. Ikeda J, Ochiai K, Morita I, Murota S (1993) Endogenous nitric oxide blocks calcium influx induced by glutamate in neurons containing NADPH diaphorase. Neurosci Lett 158: 193–196
12. Ikeda J, Ma L, Morita I, Murota S (1994) Involvement of nitric oxide and free radical (O_2^-) in neuronal injury induced by deprivation of oxygen and glucose in vitro. Acta Neurochir Suppl (wien) 60: 94–97
13. Ikeda J, Terakawa S, Murota S, Morita I, Hirakawa K (1996) Nuclear disintegration as a leading step of glutamate neurotoxicity in brain neurons. J Neurosci Res 43: 613–622
14. Kane DJ, Sarafian TA, Anton R et al (1993) Bcl-2 inhibition of neuronal death: decreased generation of reactive oxygen species. Nature 262: 1274–1277
15. Kumar S (1995) ICE-like proteases in apoptosis. Trends Biol Sci 20: 198–202
16. Kure S, Tominaga T, Yoshimoto T, Tada K, Narisawa K (1991) Glutamate triggers internucleosomal DNA cleavage in neuronal cells. Biochem Biophys Res Commun 179: 39–45
17. Macconkey DJ, Nicotera P, Hertzell P, Bellomo G, Wyllie AH, Orrenius S (1989) Glucocorticoids activate a suicide process in thymocytes through an elevation of cytosolic Ca^{2+} concentration. Arch Biochem Biophys 269: 365–370
18. Martin SJ, Green DR (1995) Protease activation during apoptosis: death by a thousand cuts? Cell 82: 349–352
19. Milligan CE, Prevette D, Yaginuma H et al (1995) Peptide inhibitors of the ICE protease family arrest programmed cell death of motoneurons in vivo and in vitro. Neuron 15: 385–393
20. Ogura A, Miyamoto M, Kudo Y (1988) Neuronal death vitro: parallelism between survivability of hippocampal neurones and sustanined elevation of cytosolic Ca^{2+} after exposure to glutamate receptor agonist. Exp Brain Res 73: 447–458
21. Ratan RR, Murphy TH, Baraban JM (1994) Oxidative stress induces apoptosis in embryonic cortical neurons. J Neurochem 62: 376–379
22. Sauer D, Allegrini PR, Cosenti A, Pataki A, Amacker H, Fagg GE (1993) Characterization of the cerebroprotective efficacy of the competitive NMDA receptor antagonist CGP40116 in a rat model of focal cerebral ischemia: an in vivo magnetic resonance imaging study. J Cereb Blood Flow Metab 13: 595–602
23. Schuier FJ, Hossmann KA (1980) Experimental brain infarcts in cats. II. Ischemic brain edema. Stroke 11: 593–601

24. Shigeno T, Yamasaki Y, Kato G et al (1990) Reduction of delayed neuronal death by inhibition of protein synthesis. Neurosci Lett 120: 117–119
25. Silos-Santiago I, Greenlund LJ, Johnson EMJ, Snider WD (1995) Molecular genetics of neuronal survival. Curr Opin Neurobiol 5: 42–49
26. Watanabe T, Yuki S, Egawa M, Nishi H (1994) Protective effects of MCI-186 on cerebral ischemia: possible involvement of free radical scavenging and antioxidant actions. J Pharmacol Exp Ther 268: 1597–1604
27. Wood KA, Youle RJ (1995) The role of free radicals and p53 in neuron apoptosis in vivo. J Neurosci 15: 5851–5857
28. Yuan J, Shaham S, Ledoux S, Eills HM, Horvitz HR (1993) The C. elegans cell death gene ced-3 encodes a protein similar to mammalian interleukin-1-β-converting enzyme. Cell 75: 641–652

II. Factors and Mechanisms Enhancing Susceptibility or Tolerance

Glutamate Enhances Brain Damage from Ischemia and Trauma*

A. Baethmann[1], F. Staub[2], O. Kempski[3], N. Plesnila[1], R.C.C. Chang[1], G.-H. Schneider[1], J. Eriskat[1], M. Stoffel[1], and F. Ringel[1]

Summary

The amino acid glutamate is a model agent to demonstrate the significance of neurotoxic mediator compounds in secondary brain damage from trauma, ischemia or other adverse conditions. Intensive research of the role of mediator compounds is clinically worthwile as more specific forms of treatment may emerge for the benefit of afflicted patients. In view of the great number of factors, cytokines, etc., which could play a role, it is mandatory that a mediator function in secondary brain damage is identified according to the stringent requirements established for that purpose. Glutamate has been shown – as is the case for only a few other agents – to meet all the criteria of a mediator of secondary brain damage. An excellent basis is thereby provided for the development of specific methods of antagonization. Although recent progress is impressive indeed, a breakthrough of the treatment of secondary brain damage in patients with head injury or cerebral ischemia by glutamate antagonists has not been accomplished so far. There may be various reasons for the delay, such as important side effects of the glutamate antagonists presently available. Some of these may be associated with inhibiting physiological transmitter functions of the amino acid. Another problem relates to the question of whether a complex disorder, such as severe head injury or stroke, can in fact be therapeutically influenced by a single compound. Severe head injury is characterized by an enormous qualitative and quantitative heterogeneity of brain lesions responsible for a clinical course which is often barely predictable. Therefore, detection of the clinical efficacy of a single pharmacological compound might be difficult, particularly in the presence of confounding factors. A trial procedure of targeting subgroups of patients could therefore be a solution. Even then the problem remains, however, that other mediator compounds, such as arachidonic acid or tissue acidosis, play a role as well which attenuate or even annihilate potential treatment effects of a glutamate antagonist. Ultimately, approaches might be required for combining various methods of antagonization which, however, should not be implemented as a rigid treatment protocol but rather as a

*Supported by Deutscher Akademischer Austauschdienst and Bundesministerium für Bildung und Forschung, FKZ: 01 KO 94026

[1] Institute of Surgical Research, Klinikum Großhadern, Ludwig-Maximilians-University, 81366 München, Germany
[2] Department of Neurosurgery, University of Cologne, Joseph Stelzmann Str. 9, 50924 Köln, Germany
[3] Institute of Neurosurgical Pathophysiology, University of Mainz, Langenbeckstr. 1, 55101 Mainz, Germany

Maturation Phenomenon in Cerebral Ischemia II
U. Ito et al. (Eds.)

flexible procedure which can be adjusted from case to case depending on the underlying dominant lesion.

Introduction

Cerebral ischemia and trauma are probably the most frequent acute threatening disorders of the brain. Both are characterized by post-insult development of secondary brain damage. Secondary brain damage summarizes the many processes of intra- and extracranial origin, which effect the course and outcome of an afflicted patient, quite likely to the same degree as the primary insult [1]. Contrary to the traumatic impact or initial interruption of blood flow, however, the various secondary sequelae are only evolving within shorter or longer time periods after the insult, which might be utilized for therapeutic intervention. Whereas many treatment modalities are obvious, such as restoration of cardiovascular and respiratory competence, removal of an intracranial mass, and lowering of intracranial pressure, other measures are less clear. These relate to the mechanisms underlying secondary brain damage, such as the development of postischemic or traumatic brain edema with opening of the blood–brain barrier, the secondary deterioration of cerebral blood flow, or the delayed demise of vulnerable nerve cells or brain tissue areas, respectively. Secondary damage evolving under these conditions is not yet fully understood, although progress must be acknowledged on various fronts, among others associated with the release and function of mediator compounds (Fig. 1), including the development of acidosis [1]. The investigation of mediator compounds of secondary brain damage is no longer simply an experimental playground, but has matured into an important clinical field of improving the outcome of cerebral trauma and ischemia.

The excitotoxic amino acid glutamate is a case in point. Abundant evidence has accumulated over the years demonstrating that specific inhibtion of the release and function of the compound affords protection in the nervous system against cerebral ischemia and trauma [6, 13, 20, 21]. Observations that antagonists remain effective, even if administered after an insult, render this approach clinically relevant and promising as it raises specificity of treatment. The requirements for identifying a mediator of secondary brain damage is perfectly illustrated for glutamate subserving the role of a model agent in this context [4]:

1. Induction of tissue or cell damage in normal brain by cerebral administration of the mediator.
2. Release and accumulation of the mediator under pathological conditions associated with the development of secondary brain damage.
3. Specific antagonization of the mediator, affording protection under pathological conditions as ischemia or trauma.

Glutamate is not only an important compound of intermediate metabolism, but also the most widespread excitatory transmitter in the brain. This notwithstanding, its release and accumulation in the extracellular space, even to levels as low as 100 μM, may wreak havoc in the adjacent tissue [25]. Its ambivalent role may explain, why glutamate was escaping recognition for a long time as one of the most destructive compounds in the brain. It was only in the late 1950s early 1960s that this role of the

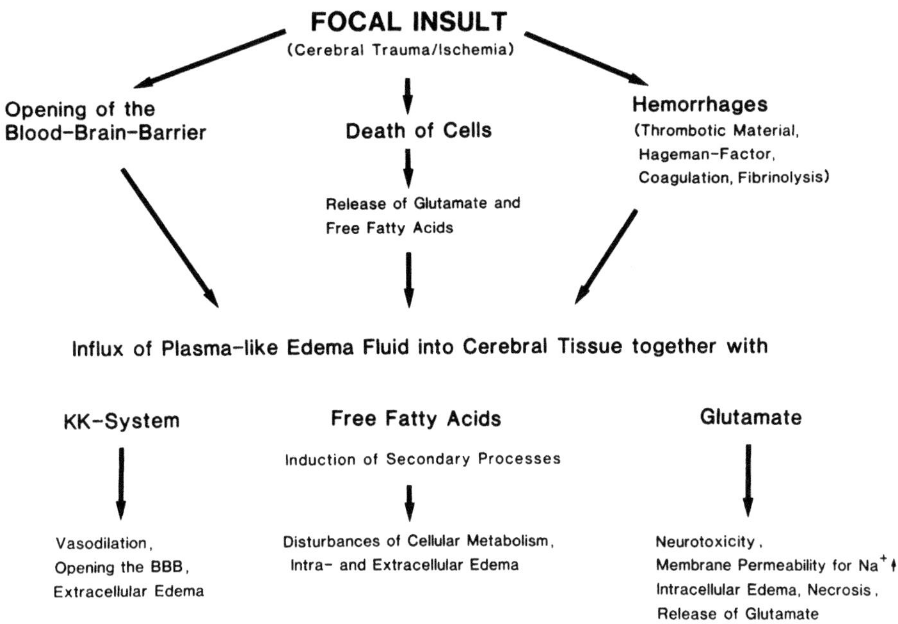

Fig. 1. The concept of release of mediator compounds in secondary brain damage, involving kinins, free-fatty acids, and glutamate in traumatic or ischemic brain injury leading to a focal lesion of irreversibly damaged parenchyma. In this area, the blood-brain barrier is opened by damage to the cerebrovascular endothelium. Furthermore, activation of lipolysis and spill-over of glutamate is facilitated by the destruction of cells occurring in the primary lesion. Induction of coagulation taking place in the focus is a potent stimulus for activation of the kallikrein-kinin system, which adversely influences the blood-brain barrier function and microcirculation. Altogether, release and accumulation of kinins, of arachidonic acid, and of glutamate must be considered to enhance the primary insult by various mechanisms, e.g., supporting formation of perifocal brain edema, disturbances of microcirculation, and – last but not least – cytotoxicity. The spread of irreversibly damaged parenchyma evolving under these circumstances with a delay might be attributed to the release and accumulation of mediator compounds into the penumbra zone [3]

amino acid was identified in the pioneering studies of Lucas and Newhouse [17], Van Harreveld [31], and Olney and Sharpe [19].

Glutamate as Mediator of Secondary Brain Damage

Van Harreveld conducted experiments using brain tissue extracts, which enabled him to suggest that the amino acid glutamate plays a role in the development of cell swelling which occurs during cerebral ischemia or spreading depression [31]. The conclusions were based on observations of glutamate to increase Na^+ membrane permeability, resulting in a shift of extracellular fluid into the intracellular compartment. The enormous destructive potential of glutamate was directly demonstrated in experiments on cerebral administration of the amino acid to the brain in vivo using iontophoresis [30]. The procedure led to a halo of completely extinct brain cortex characterized by ballooning of the tissue elements including glial and nerve cell processes. Other approaches were pursued by Lucas and Newhouse [17] and Olney and Sharpe [19], who observed degenerative changes in the brain or retina after systemic administration. The intracellular fluid shift, such as occurs in complete ischemia of the brain, was explained as being brought about by a release of glutamate from the intracellular compartment [29]. Van Harreveld was also postulating that spreading depression, secondary to a mechanical stimulus or focal administration of high K^+ levels, is elicited by a release of glutamate; however, not of K^+-ions as assumed by Grafstein [10].

This laboratory has carried out a series of in vivo and in vitro studies in order to explore the role of glutamic acid in the formation of brain edema [15, 24, 26]. By using different experimental models, such as ventriculo-cisternal perfusion of the brain with glutamate, it was found that the cerebral water and Na^+ content was markedly increased while the cerebral K^+ content decreased, indicating intracellular fluid accumulation characteristic of cytotoxic brain edema [24]. This conclusion was supported by analysis of the specific electrical resistance (impedance) of the brain during cerebral administration of glutamate. Cerebral administration of glutamate was associated with a significant decrease in electrical conductivity of the brain (or increase in impedance) from shrinking of the extracellular compartment secondary to translocation of fluid into the cells [15]. The results confirm electron-microscopic studies of spreading depression or circulatory arrest, using a specific fixation procedure (freeze substitution) to preserve the fluid distribution between the intra- and extracellular compartment [29]. Taken together, these data unanimously confirm that glutamate induces brain damage following its administration to intact brain.

Induction of Cell Swelling by Glutamate

Attempts were also made to study cell swelling from glutamate in vitro, using tissue slices or suspended cells [9, 14, 26]. This approach has a variety of advantages, particularly of better control and adjustment of ambient conditions, such as pH, PO_2, temperature, electrolyte levels, osmolarity, etc., facilitating assessment of cause–effect relationships which, under pathological conditions, is more difficult in vivo in an organ as complex as the brain. When utilizing stable cell lines such as C6 glioma, the temporal course of cell swelling from glutamate, including dose-response characteristics, can be quantitatively analyzed at an excellent resolution. Moreover, underlying cell biological mechanisms can also be explored. In studies using suspensions of C6 glioma an increase in cell volume was already observed at a glutamate concentration in the medium of as low as 50 μM. The swelling response, however, had intriguing kinetics, with the maximum cell size obtained at 30–40 min after the addition of glutamate, followed by stepwise normalization. When, however, critical glutamate levels were added to the cell suspension (>200 μM), cell volume recovery was incomplete or even failing [26]. Former measurements of the O_2 uptake of the glial cells had revealed a marked enhancement of cell respiration by glutamate [16]. The phenomenon of spontanous cell volume recovery was investigated by assessing the glutamate concentration following addition of the compound to the cell suspension. The clearance of the compound from the medium thereby observed provided important evidence. Cell volume recovery was found to commence as soon as the glutamate level in the suspension was decreased to a sub-threshold level. It was concluded that the decline of glutamate in the medium following its administration was brought about by glial uptake underlying stimulation of the cellular O_2 consumption.

It is known that astrocytes have high-affinity uptake mechanisms (transporter) for glutamate [27], probably also glioma cells, which were most likely activated during the cell swelling experiments. In fact, uptake and accumulation of glutamate, which is occurring against a steep concentration gradient, involves co-uptake of two to three Na^+ ions per glutamate molecule as an immediate source of energy. Clearance of glutamate can thus be viewed as the mechanism underlying cell swelling from a net increase of intracellular osmolarity. The spontaneous cell volume recovery after induction of the glutamate-induced cell swelling, as observed in these studies [26], would require that the excessive osmotic load is subsequently removed from the cells, e.g., by activation of the Na^+/K^+-adenosine triphosphatase (ATPase). This would represent no problem with a normal substrate- and O_2-supply in vitro or in vivo. Respective conditions, however, are most likely impaired in ischemic brain tissue or after severe head injury, for example in the perifocal penumbra zone suffering from an unreliable tissue perfusion. Under these circumstances, the glial clearance of excessive glutamate together with Na^+ ions from the extracellular space, e.g., in the vicinity of a necrosis, would rapidly exhaust the remaining energy-rich phosphates, thereby preventing elimination of Na^+ ions accumulating in the cells. Consequently, the glial cells which are swollen in fulfilment of their function to reestablish the extracellular homeostasis can not normalize their increased cell volume due to energy failure. Hence, the initially compensatory cell volume increase would change into a pathological phenomenon – cytotoxic cell swelling.

The above hypothesis that the glutamate-induced cell swelling is a compensatory response resulting from uptake of glutamate and Na^+ ions, eventually activating Na^+/K^+-ATPase for removal of the intracellularly accumulating Na^+ ions, was tested in additional experiments using ouabain. Briefly, cell swelling was more or less completely inhibited when the glial cells were incubated with the drug prior to administration of glutamate [26]. Obviously, ouabain was paralysing removal of intracellular-accumulating Na^+ ions by the Na^+/K^+-ATPase, flattening the extra- to intracellular Na^+-concentration gradient which is required to fuel the cellular uptake of glutamate. Therefore, the downhill influx of Na^+ ions together with glutamate was annihilated, preventing accumulation of the amino acid in the cells, and thus cell swelling. This experimental finding may reflect the situation in brain tissue with energy failure from cerebral ischemia leading to the disappearance of the transmembrane Na^+ ion concentration gradient. Then, glutamate released into the interstitial space can not be cleared by the glial cells, resulting in its accumulation, as frequently confirmed by microdialysis or interstitial fluid drainage of the brain under pathological conditions (see below).

Release of Glutamate in Traumatic Brain Edema

The requirement that a mediator compound should accumulate in pathological concentrations in the brain (or a compartment where it is normally absent) is also satisfied by glutamate [2, 3]. The amino acid occurs in normal brain tissue at concentrations of approximately 10 m*M*; however, this is almost exclusively in the intracellular compartment, where it may reach even higher concentrations with regard to the low extracellular level. Consequently, the extra-to-intracellular glutamate concentration gradient may approach 1:1000 to 1:1500. Therefore, diffuse cell destruction in the brain, as in infarction or traumatic contusion, can be considered to facilitate spillover of glutamate into the extracellular space of the penumbra zone surrounding the lesion. An acute decrease of blood flow below a critical level ("infarction threshold") is characterized by a widespread terminal depolarization of the cells in association with a marked release of glutamate [28] together with K^+ ions. The extracellular increase of glutamate may, therefore, be attributed to a variety of mechanisms, such as: (a) diffusion from necrotic tissue; (b) ischemic release from the intracellular compartment, together with (c) a failing clearance by the glial cell elements. This scenario is confirmed by many experimental, and recently also clinical studies, analyzing the release of the amino acid into the interstitial compartment of the brain suffering from ischemia or trauma (see below).

This laboratory has carried out investigations on the release of glutamate from a focal lesion of the brain (cold injury) in experimental animals using interstitial fluid drainage. For identification of vasogenic edema fluid harvested by this method, the animals were injected i.v. with Evans blue as a blood-brain barrier marker. The clear vasogenic fluid collected was not contaminated by either tissue or blood and, therefore, could be utilized for the assessment of glutamate and other mediator compounds, such as arachidonic acid. The focal trauma led to a marked increase of the glutamate level in the edema fluid to approximately ten times the concentration found in normal cerebrospinal fluid (CSF) utilized as a reference [2]. The effect of plasma-borne glutamate was studied by the measurement of plasma samples. Accordingly, the edema fluid had a glutamate level which was exceeding the concurrent plasma concentration by two to three times, confirming the major contribution of the release of glutamate from brain tissue proper as compared to the influx of the amino acid through the broken blood-brain barrier into the brain.

The experiments further highlighted a close association between the release of glutamate and the severity of the insult. In a subgroup of animals (approximately 50% of the total group) the intracranial pressure was vigorously increasing from the evolving brain edema to a level which was eventually interfering with the maintenance of cerebral blood flow. This became apparent by the development of an isoelectric electroencephalogram (EEG) and respective changes of energy-rich phosphate compounds (ATP, creatine phosphate). Such a course led to a significantly more pronounced release of glutamate into the interstitial edema fluid, reaching no less than 100–150 times the concentration found in normal CSF [2].

Release and accumulation of glutamate in the brain has been confirmed in many studies of cerebral ischemia [5, 11], but also in trauma, such as fluid percussion injury [12]. Moreover, accumulation of glutamate in cerebral microdialysate has been confirmed, meanwhile, in patients with severe head injury [8, 23]. Thus it was demonstrated that the release of the amino acid occurs in association with an adverse clinical course such as intracranial hypertension. Based on this experience it might be expected that microdialysis of human brain will eventually mature to a reliable method of "chemical" monitoring as an early warning procedure against the development of serious complications.

Inhibition of Secondary Brain Damage by Glutamate Antagonization

A great number of studies have been carried out to analyze methods of inhibition of the fatal sequelae of the release of glutamate in the brain. This is markedly facilitated by the availability of specific, or non specific glutamate receptor antagonists, such as MK 801 (dizocilpine), the more recently available α-amino-3-hydroxy-5-methylisoxazole-4-propionic acid (AMPA) receptor blocking agent 2,3-dihydroxy-6-nitro-7-sulfamoylbenzo (F) quinoxaline (NBQX), or the specific *N*-methyl-D-asparate-(NMDA) receptor blocker CGS 19755, which seems to be able to permeate the intact blood-brain barrier [6, 20–22]. As far as the NMDA-receptor-dependent channel blocker MK 801 is concerned, data on limiting ischemic infarction following middle cerebral artery (MCA) occlusion were highly impressive [20, 21], whereas the agent was observed to fail in global cerebral ischemia [7]. Antagonists such as the AMPA receptor blocker NBQX, appear to be clinically useful since they have a therapeutic efficacy even when administered after the insult [6]. These results demonstrate that secondary brain damage progresses after the initial insult for a considerable time period, making it worthwhile to devise effective procedures of antagonization as long as the "window of opportunity" is open.

Respective clinical consequences have already been drawn. Meanwhile, glutamate antagonists are being tested in various clinical trials, such as in head injury or subarachnoid hemorrhage. Nevertheless, as in the case of corticosteroids, expectations concerning the outcome should be restrained in view of the enormous quantitative and qualitative heterogeneity of severe head injury in patients. This complex disorder is characterized by lesions of differing severity, which may be variably (or not at all) responsive to treatment with glutamate antagonists. A trial design lumping all cases with lesions of different severity and nature in one group might be bound to fail, if only a subgroup of lesions is drug-responsive. Whether a more target-oriented trial design provides for a better solution, however, remains to be seen. These problems notwithstanding, the emergence of a convincing therapeutic effect would render glutamate antagonists the first agents available for specific medical treatment of secondary brain damage in severe head injury, a condition which so far is subject only to measures of reestablishing or maintaining cerebral blood flow and O_2 supply, and to lowering intracranial pressure [18].

Acknowledgments. The secretarial assistance of Helga Kleylein and Edith Martin is gratefully acknowledged.

References

1. Baethmann A, Kempski O, Schürer L (eds) (1993) Mechanisms of secondary brain damage – current state. Acta Neurochirg, Suppl. 57, Springer, Vienna New York

2. Baethmann A, Maier-Hauff K, Schürer L, Lange M, Guggenbichler C, Vogt W, Jacob K, Kempski O (1989) Release of glutamate and of free-fatty acids in vasogenic brain edema. J Neurosurg 70: 578–591
3. Baethmann A, Maier-Hauff K, Kempski O, Unterberg A, Wahl M, Schürer L (1988) Mediators of brain edema and secondary brain damage. Crit Care Med 16: 972–978
4. Baethmann A (1978) Pathophysiological and pathochemical aspects of cerebral edema. Neurosurg Rev 1: 85–100
5. Benveniste H, Drejer J, Schousboe A, Diemer NH (1984) Elevation of the extracellular concentrations of glutamate and aspartate in rat hippocampus during transient cerebral ischemia monitored by intracerebral microdialysis. J Neurochem 43: 1369–1374
6. Buchan AM, Xue D, Huang ZG, Smith KH, Lesiuk H (1991) Delayed AMPA receptor blockade reduces cerebral infarction induced by focal ischemia. Neuroreport 2: 473–476
7. Buchan AM, Li H, Pulsinelli WA (1991) The *N*-methyl-D-aspartate antagonist, MK-801, fails to protect against neuronal damage caused by transient, severe forebrain ischemia in adult rats. J Neurosci 11: 1049–1056
8. Bullock R, Zauner A, Woodward J, Marmarou A, Young H (1995) Excitatory amino acid release patterns after severe human head injury – experience with microdialysis in 30 patients. J Neurotrauma 12: 372-410
9. Chan PH, Fishman RA, Lee JL, Candelise L (1979) Effects of excitatory neurotransmitter amino acids on swelling of rat brain cortical slices. J Neurochem 33: 1309–1315
10. Grafstein B (1956) Mechanism of spreading cortical depression. J Neurophysiol 19: 154–171
11. Hagberg H, Lehmann A, Sandberg M, Nyström B, Jacobson I, Hamberger A (1985) Ischemia-induced shift of inhibitory and excitatory amino acids from intra- to extracellular compartments. J Cereb Blood Flow Metabol 5: 413–419
12. Katayama Y, Becker DP, Tamura T, Hovda DA (1990) Massive increases in extracellular potassium and the indiscriminate release of glutamate following concussive brain injury. J Neurosurg 73: 889–900
13. Kempski O (ed) (1993) Glutamate – transmitter and toxin. Zuckschwerdt, Munich (Klinische Ernährung, vol 36)
14. Kempski O, Staub F, Schneider GH, Weigt GH, Baethmann A (1992) Swelling of C6 glioma cells and astrocytes from glutamate, high K^+ concentration or acidosis. Prog Brain Res 94: 69–75
15. Kempski O (1982) Die Lokalisation des Glutamat-induzierten Hirnödems. Thesis, Ludwig-Maximilians-University, Munich
16. Kempski O, Gross U, Baethmann A (1982) An in vitro model of cytotoxic brain edema: cell volume and metabolism of cultivated glial and nerve cells. Adv Neurosurg 10: 254–258
17. Lucas DR, Newhouse JP (1957) The toxic effect of sodium L-glutamate on the inner layers of the retina. AMA Arch Ophthalmol 58: 193–201
18. Miller JD (1992) Evaluation and treatment of head injury in adults. Neurosurg Quart 2: 28–43
19. Olney JW, Sharpe LG (1969) Brain lesions in an infant rhesus monkey treated with monosodium glutamate. Science 166: 386–388
20. Ozyurt E, Graham DI, Woodruff GN, McCulloch J (1988) Protective effect of the glutamate antagonist MK-801 in focal cerebral ischemia in the cat. J Cereb Blood Flow Metab 8: 138–143
21. Park CK, Nehls DG, Graham DI, Teasdale GM, McCulloch J (1988) The glutamate antagonist MK-801 reduces focal ischemic brain damage in the rat. Ann Neurol 24: 543–551
22. Perez-Pinzon MA, Maier CM, Yoon EJ, Sun GH, Giffard RG, Steinberg GK (1995) Correlation of CGS 19755 neuroprotection against in vitro excitotoxicity and focal cerebral ischemia. J Cereb Blood Flow Metabol 15: 865–876
23. Persson L, Hillered L (1992) Chemical monitoring of neurosurgical intensive care patients using intracerebral microdialysis. J Neurosurg 76: 72–80
24. Rothenfußer W (1982) Die Bedeutung von Glutamat als Hirnödemfaktor. Thesis, Ludwig-Maximilians-University, Munich
25. Rothman SM, Olney JW (1986) Glutamate and the pathophysiology of hypoxic-ischemic brain damage. Ann Neurol 19: 105–111
26. Schneider GH, Baethmann A, Kempski O (1992) Mechanisms of glial swelling induced by glutamate. Can J Physiol Pharmacol 70: S334–S343
27. Schousboe A (1981) Transport and metabolism of glutamate and GABA in neurons and glial cells. Int Rev Neurobiol 22: 1–45

28. Shimada N, Graf R, Rosner G, Wakayama A, George CP, Heiss WD (1989) Ischemic flow threshold for extracellular glutamate increase in cat cortex. J Cereb Blood Flow Metabol 9: 603–606
29. Van Harreveld A (1972) The extracellular space in the vertebrate central nervous system. In: Bourne GH (ed) The structure and function of nervous tissue IV. Academic Press, New York, pp 447–511
30. Van Harreveld A, Fifkova E (1971) Light- and electron-microscopic changes in central nervous tissue after electrophoretic injection of glutamate. Exp Mol Pathol 15: 61–81
31. Van Harreveld A (1959) Compounds in brain extracts causing spreading depression of cerebral cortical activity and contraction of crustacean muscle. J Neurochem 3: 300–315

Glutamate-Mediated Mechanisms in Delayed Neuronal Death After Cerebral Ischemia

N.H. Diemer, T. Bruhn, T. Christensen, M. Nielsen, and F.F. Johansen

Summary

The density of ionotropic glutamate receptors on the ischemia-vulnerable CA1 pyramidal cells is one of the highest in the CNS. Postischemic use of the AMPA antagonist NBQX has shown up to 80% protection against cell death. Three groups of rats given NBQX (30 mg/kg × 3) immediately after ischemia and with a survival period of 6 days, 3 weeks, and 8 weeks were compared to untreated groups of ischemic rats. Countings of CA1 pyramidal cells showed an equal, significant protection in all three groups (20%–40% cell loss). Thus, if glutamate antagonist treatment is instituted immediately after ischemia, the obtained protection is permanent. Others [11] have shown that a 6-h delay of treatment results in loss of protection. The very delayed loss in the latter situation could be a variant of the maturation phenomeon described by Ito and colleague (1975) [9].

Introduction

A short period of global cerebral ischemia in rodents induces loss of hippocampal pyramidal CA1 neurons. A number of pharmacological treatments as well as hypothermia have been evaluated, but until now glutamate antagonists of the amino-3-hydroxy-5-methyl-4-isoxazoleprionic acid (AMPA) type have turned out to protect most effectively. Since the prevalent excitatory receptors on the ischemia-vulnerable CA1 neurons (as well as in a number of other vulnerable neuron populations) are of the glutamate type, this can explain why AMPA antagonists are so effective in protecting these particular neurons [5, 19]. However, in most studies of glutamate antagonists the survival period has seldom exceeded 1 week. Recently a study by Li and Buchan (1995) indicated that the protective effect of the AMPA antagonist 2,3-dihydroxy-6-nitro-7-sulfamoylbenzo(F)quinoxaline (NBQX) is present after 1 week of survival but has disappeared after 4 weeks [11]. The drug was given after 6 h of recirculation; however, it is known that administration of an AMPA antagonist is most effective when treatment is started immediately after ischemia (Diemer et al. 1996). The present investigation studied the long-term effect of immediately postischemic administration of NBQX in groups of rats with 1, 3, and 8 weeks of survival.

Institute of Neuropathology, University of Copenhagen, Denmark

Maturation Phenomenon in Cerebral Ischemia II
U. Ito et al. (Eds.)

Material and Methods

Male Wistar rats (250–300 g) were subjected to transient forebrain ischemia using a modification of Pulsinelli and Brierley's four-vessel occlusion model. Three groups of drug-treated animals were established, all given NBQX (30 mg/kg × 3 i.p.) immediately after ischemia with 6-day survival, 3-week survival, and 8-week survival ($n = 8$, each group). In addition, there were three groups of Vehicle-treated rats, with 6-day, 3-week, and 8-week survival ($n = 8$, each group).

Both vertebral arteries were electrocoagulated in methohexital (50 mg/kg i.p.) anesthesia. The rats were allowed to recover with free access to water but were fasted overnight. On the following day the rats were anesthetized with 1% halothane in a 2:1 N_2O/O_2 mixture, intubated, and mechanically normoventilated in a rodent respirator (New England Medical Instruments). The rats had a femoral artery cannulated to record mean arterial blood pressure and to obtain blood samples for measurements of PO_2, PCO_2, pH, and plasma glucose concentration (ABL 2 acid-base laboratory, Radiometer, Copenhagen, and Beckman Glucose Analyzer 2). The common carotid arteries were gently exposed, and 3-0 silk ligatures threaded through polyethylene tubing were placed around them. The halothane was turned off, and before the rats recovered from anesthesia the carotids were occluded by tightening the ligatures. Blood was withdrawn from the femoral artery until the pupils dilated and became unresponsive to light, while keeping the mean arterial blood pressure at 50 or 60 mmHg. The body and head (temporal muscle) temperature was kept at 37°C–37.5°C and 36°C, respectively. After 12 min the ligatures were released and the blood reinfused. The animals were allowed to survive according to the schedule shown above.

The perfusion fixed brains were cut subserially in 7-μm frontal sections. Determination of cell density in the intermediate zone of the dorsal hippocampus was performed using an ocular grid and ×40 magnification on four sections. Only neurons with a distinct nuclear membrane and basophilic cytoplasm were counted. Astrocyte and microglia nuclei could easily be distinguished from neuron nuclei.

Statistical analysis was performed using Kruskal-Wallis one-way analysis of variance by ranks, and a significance level of 0.05 was chosen.

Results

The median values of the quantitations of CA1 cells in the dorsal hippocampus are summarized in Table 1. Untreated rats with 6-day survival had a loss of 88% pyramidal neurons. NBQX-treated rats with 6-day survival had only 19% loss ($p < 0.01$). In the groups with 3-week survival the untreated median value was 79% loss whereas the NBQX animals had 25% loss ($p < 0.05$). Also in the 8-week groups the difference was significant ($p < 0.02$). Here the median values of loss were 70% and 22%, respectively. No frank infarctions were seen, and there were no bleedings. In the long-term surviving groups, especially at 8 weeks, there was beginning shrinkage of the CA1 sector. All neurons with irreversible nuclear changes indicating cell death (karyorhhexis, shrunken triangular nuclei) had distinctly eosinophilic cytoplasm. Apoptotic bodies were not seen.

Table 1. Data on remaining CA1 neurons in dorsal hippocampus in the various groups; medians with range (normal number of CA1 pyramidal neurons per field is 53, range 48–55)

Survival period	6 days	3 weeks	8 weeks
Ischemia (min)	12	12	12
Hypotension (mmHg)	50	50	60
Untreated, % of CA1 pyramidal neurons	12 (5–48)	21 (12–55)	30 (10–70)
NBQX, % of CA1 pyramidal neurons	81 (55–95)	75 (53–82)	78 (56–90)
p (Kruskal-Wallis test)	$p < 0.001$	$p < 0.05$	$p < 0.02$

Discussion

AMPA Antagonist Treatment

In contrast to the striking effect of NMDA antagonists in reducing infarct size after middle cerebral artery occlusion [8, 12]. On the other hand, AMPA antagonists have been shown to be very effective in protecting against CA1 pyramidal cell loss, especially after shorter periods of ischemia [5, 14]. It cannot be excluded that at least part of the effect is due to a depression of normo- or hypermetabolism in hippocampus [13, 15]. The intracellular mechanisms of AMPA antagonist protection against ischemia-induced damage have not been discovered, but an unspecific reduction of the excitation/inhibition ratio leading to reduced energy metabolism stress is a likely possibility. Other classes of compounds have also shown (somewhat less) protective effects which can be explained by such a common mechanism. Since some of the most sensitive neuron types (including Purkinje cells) also have a high density of glutamate receptors, this might explain the protective effect of, for example, AMPA antagonists after global ischemia. NBQX, which is a competitive AMPA antagonist has been found to induce long-term effects on metabolic processes such as protein synthesis [7].

Glutamate-Mediated Mechanisms and the Maturation Phenomenon: Long-Term Effect of NBQX Treatment

In 1975 Ito et al. [9] described the very important maturation phenomenon. It was originally discovered in hippocampus, but it is probably also part of the ischemia-induced damage in other neuronal populations. Recent studies on the effect of protective paradigms have shown that a very slow maturation takes place [2, 6, 11] after what was initially thought to be effective "anti-ischemic" treatment by means of hypothermia [1, 3] or AMPA antagonists. The present study showed the same degree of protection in rats which survived for up to 8 weeks after 12 min of four-vessel occlusion as in those surviving for 6 days. In a study with 7 or 28 days survival after 10 min of four-vessel occlusion and NBQX treatment 6 h postischemically there was 45% injured CA1 neurons on day 7 and 89% injured neurons on day 28 [11]. This significant difference indicates that with this paradigm the NBQX-induced protection is not permanent. Thus a 6-h delay period before treatment is started results in a quite different final outcome. It is not known how slowly the neurons disappear in the case of insufficient protection or whether apoptotic mechanisms are operating. Light microscopic morphology, despite its lack of precision, shows no new patterns of nuclear or cytoplasmic degeneration, but it is obvious that the paradigm of delayed insufficient protection is interesting in itself as a model of subacute neurodegeneration.

Acknowledgments. This study was supported by grants from the Danish Health Science Council and the Danish State Biotechnological Programme (PharmaBiotec). Eva Rahtkens is thanked for excellent technical support.

References

1. Busto R, Dietrich WD, Globus MY-T, Ginsberg MD (1989) Postischemic moderate hypothermia inhibits CA1 hippocampal ischemic neuronal injury. Neurosic Lett 101: 299–304
2. Colbourne F, Corbett D (1994) Delayed and prolonged post-ischemic hypothermia is neuroprotective in the gerbil. Brain Res 654: 265–272
3. Coimbra C, Wieloch T (1994) Moderate hypothermia mitigates neuronal damage in the rat brain when initiated several hours following transient cerebral ischemia. Acta Neuropathol (Berl) 87: 325–331
4. Diemer NH, Siemkowicz E (1982) Increased 2-deoxyglucose uptake in hippocampus, substantia nigra and globus pallidus after cerebral ischemia in the rat. Acta Neurol Scand 61: 56–63
5. Diemer NH, Johansen FF, Jørgensen MB (1990) *N*-Methyl-D-aspartate and non-*N*-Methyl-D-aspartate antagonists in global cerebral ischemia. Stroke 21: III, 39–42
6. Dietrich WD, Busto R, Alonso O, Globus MY-T, Ginsberg MD (1993) Intraischemic but not postischemic brain hypothermia protects chronically following global forebrain ischemia in rats. J Cereb Blood Flow Metab 13: 541–549
7. Frank L, Bruhn T, Diemer NH (1992) Regional cerebral protein synthesis after transient ischemia in the rat: effect of the AMPA antagonist NBQX. Neurosci Lett 140: 108–112
8. Gill R, Brazell C, Woodruf GN (1988) The neuroprotective action of dizicilpine (MK-801) in the rat middle cerebral artery occlusion model of focal ischemia. Br J Pharmacol 103: 2030–2036
9. Ito U, Spatz M, Walker JT, Klatzo I (1975) Experimental cerebral ischemia in Mongolian gerbils. Acta Neuropathol (Berl) 32: 209–223
10. Kirino T (1982) Delayed neuronal death in the gerbil hippocampus following ischemia. Brain Res 239: 57–69
11. Li H, Buchan AM (1995) Progressive loss of hippocampal CA1 neurons following transient forebrain ischemia. J Cereb Blood Flow Metab 15 [suppl 1]: S246
12. Mies G, Kohno K, Hossmann KA (1993) MK-801, a glutamate antagonist, lowers flow threshold for inhibition of protein synthesis after middle cerebral artery occlusion of rat. Neurosci Lett 155: 65–68
13. Pulsinelli WA, Levy DE, Duffy TE (1982) Regional cerebral blood flow and glucose metabolism following transient forebrain ischemia. Ann Neurol 11: 499–509
14. Sheardown MJ, Nielsen EØ, Hansen AJ, Jacobsen P, Honoré T (1990) 2,3-Dihydroxy-6-nitro-7-sulfamoylbenzo(F)quinoxaline: a neuroprotectant for cerebral ischemia. Science 247: 571–574
15. Suzuki R, Yamaguchi T, Li C-L, Klatzo I (1983) The effects of 5-minute ischemia in Mongolian gerbils. II. Changes in spontaneous activity in cerebral cortex and CA1 sector of hippocampus. Acta Neuropathol 60: 217–222

Neuroprotection by Drug-Induced Neurotrophic Factors

J. Krieglstein, I. Semkova, C. Culmsee, and P. Wolz

Summary

Neurotrophic factors are of great interest for the therapy of neurodegenerative diseases. However, they are proteins which scarcely permeate across the blood–brain barrier (BBB). An elegant approach to increasing the availability of neurotrophic factors in the central nervous system (CNS) might be to induce the synthesis and release of these proteins in brain tissue by lipophilic drugs. To examine this issue we have selected clenbuterol, a β_2-adrenergic receptor agonist, and selegiline, an irreversible monoaminoxidase (MAO)-B inhibitor, both shown to enhance nerve growth factor (NGF) synthesis. We tried to find out whether clenbuterol and selegiline can rescue cultured hippocampal neurons from excitotoxic damage by induction of NGF. Additionally, in the rat model of focal cerebral ischemia the ability of these compounds to reduce the infarct volume was examined.

Exposure of mixed neuronal/glial hippocampal cultures to clenbuterol (1–100 μM) and selegiline (10 pM–1 nM) significantly enhanced the content of NGF measured in the culture medium by two-site enzyme linked immunosorbent assay (ELISA). The excitotoxic injury was induced in the same type of cells after 14 days in vitro by exposure to 0.5 or 1 mM L-glutamate for 1 h in serum-free medium. Clenbuterol (1–100 μM) and selegiline (10 pM–1 nM) were able to protect hippocampal neurons from glutamate-induced damage. Furthermore, both clenbuterol (0.1 and 0.5 mg/kg) and selegiline (8 × 15 mg/kg) intraperitoneally administered significantly reduced the cortical infarct volume in the rat model of focal cerebral ischemia.

The present data demonstrate that clenbuterol and selegiline induce NGF synthesis in cultured hippocampal cells and protect hippocampal neurons from excitotoxic damage. The neuroprotective activity of these drugs was also demonstrated in vivo in a rat model of cerebral ischemia. The results offer strong evidence that pharmacologically induced NGF can provide neuroprotection against ischemic injury to CNS.

Introduction

An "excitotoxic" form of neuronal injury and death has been postulated to underlie the neurodegeneration that occurs after ischemic insults and trauma [8, 9, 40]. Addi-

Institut für Pharmakologie und Toxikologie, Fachbereich Pharmazie und Lebensmittelchemie, Philipps-Universität, Ketzerbach 63, 35032 Marburg, Germany

Maturation Phenomenon in Cerebral Ischemia II
U. Ito et al. (Eds.)

tionally, an excitotoxic component has been shown to play an important role in the pathogenesis of chronic neurodegenerative disorders such as Alzheimer's disease, Parkinson's disease, amyotrophic lateral sclerosis and Huntington's disease, which are characterized by progressive loss of neuronal elements [15, 27, 28, 30]. This cell loss may be prevented by the localized delivery of neurotrophic factors. Among other neurotrophic substances nerve growth factor (NGF) has been characterized as a survival and growth-promoting protein for the cholinergic neurons of the basal forebrain [17, 43], but the effects of NGF are probably not restricted to these neuronal populations [38]. Recently, it has been shown that NGF protects rat hippocampus from ischemic insult [5, 29, 39] and attenuates metabolic–excitotoxic and iron-induced degeneration of cultured hippocampal neurons [6, 7, 25, 45]. This suggests that NGF may have protective functions on this brain region which is particularly vulnerable to cerebral ischemia and chronic neurodegenerative disorders (e.g., Alzheimer's disease). Since NGF is unable to penetrate the blood–brain barrier (BBB), understanding the mechanisms which regulate its endogenous expression in the brain could be useful for the development of pharmacological therapy of both acute brain injury and chronic neurodegenerative diseases associated with NGF deficiency. By pharmacological manipulation of endogenous regulatory systems the synthesis of NGF might be elevated and/or sustained during the critical period of injury. Thus, induction of NGF by pharmacological intervention may ameliorate the extent of neuronal damage after acute events such as stroke and trauma and may have therapeutic benefits for Alzheimer's disease.

To examine this issue we have selected clenbuterol, a lipohilic β_2-adrenergic receptor agonist which has been shown to increase NGF synthesis in intact rat brain [12, 18], and selegiline, an irreversible monoaminoxidase (MAO)-B inhibitor shown to induce NGF synthesis in cultured rat cortical astrocytes [36]. Therefore, we have tried to find out whether clenbuterol and selegiline can rescue cultured hippocampal neurons from excitotoxic damage by induction of NGF. Additionally, the ability of these compounds to reduce neuronal degeneration in vivo was tested in a rat model of focal cerebral ischemia.

Materials and Methods

Animals

Neonatal (P1) Fischer 344 rats and Long Evans rats (225–275 g, Møllegaard, Denmark) were used. The animals had free access to food (Altromin, Lage, Germany) and water and were kept under environmental standardized conditions (12 h dark–light cycle, 23°C ± 1°C and 55% relative humidity).

Primary Rat Hippocampal Cultures

Mixed neuronal–glial primary hippocampal cultures were prepared from neonatal (P1) Fischer 344 rats. The animals were decapitated under sterile conditions and the hippocampi were isolated and incubated for 20 min at 37°C in Leibovitz's L-15 medium supplemented with 1 mg/ml papain and 0.2 mg/ml bovine serum albumin. The supernatant was removed and the tissue pieces were triturated in minimum essential Eagle's medium (MEM) (Gibco BRL, Scotland) containing 10% NU serum. The supernatant was transferred to another tube and the undispersed pieces were triturated again through a fire-polished pasteur pipette. The supernatants were combined and trypsin inhibitor solution (1% in MEM containing 10 mg/ml bovine serum albumin) was layered under the cell suspension. The mixture was centrifuged at 600 rpm and 20°C for 10 min. The cells were resuspended in MEM containing 10% NU serum and seeded on poly-L-lysine-coated 35-mm culture dishes (3×10^7 cells/dish).

The cells were cultivated in MEM supplemented with 10% NU serum, 20 units/ml penicillin, 20 μg/ml streptomycin, and were kept in a humidified atmosphere of 95% air and 5% CO_2 at 37°C. After 2 days in culture cytosine-β-D-arabinofuranoside (1 μM) was added to the medium for 24–36 h to minimize glial cell proliferation. The medium was changed every 3–4 days.

Two-Site Enzyme Immunoassay for Nerve Growth Factor

NGF content was measured in the medium of primary rat hippocampal cultures. The cells were grown for 14 days and then exposed to different concentrations of clenbuterol (0.1–100 μM) for 4 h and selegiline (1 pM to 100 nM) for 6 h in serum-free medium. The control sister cultures received the same volume of saline. The samples were centrifuged to remove the cell debris. After aprotinin (20 μg/ml) was added the supernatants were stored at –20°C until use. An enzyme linked immunosorbent assay (ELISA) for NGF was performed as described by Korshing and Thoenen [21] with minor modifications. Standard curves were prepared with 2.5 S NGF-β standard (Boehringer, Mannheim) purified from mouse submaxillary glands by assaying parallel wells containing increasing amounts of NGF-β. The amount of NGF-β in the samples was calculated from the standard curve.

Glutamate Neurotoxicity in Rat Hippocampal Cultures

Rat hippocampal neurons were damaged with L-glutamate after 14 days in vitro as described previously [20]. The cultures were washed with serum-free MEM and then exposed to serum-free MEM containing 0.5 mM or 1 mM L-glutamate for 1 h. Afterwards, the glutamate was washed away twice with serum-free MEM and then fresh serum-free MEM was added to the cultures for 18 h (the point of measurement of glutamate toxicity). Cultures that were not exposed to L-glutamate were also washed with serum-free MEM and switched to serum-free MEM without glutamate. The percentage of damaged neurons was determined 18 h later by trypan blue exclusion. Cells stained with trypan blue were regarded as non-viable.

The effects of clenbuterol (0.1–100 μM) and selegiline (1 pM–10 nM) on glutamate-induced neurodegeneration were investigated. The drugs were present in the medium 4 h (clenbuterol) and 6 h (selegiline) before excitotoxic injury and until 18 h after this injury, including the period of exposure to glutamate. Control cultures received the vehicle only.

Permanent Focal Cerebral Ischemia in Rats

Permanent middle cerebral artery (MCA) occlusion was performed in male Long Evans rats as described by Tamura et al. [41] with minor modifications. Animals were anesthetized with a mixture of 68.5% N_2O/30% O_2 that contained 1.5% halothane. A vertical incision was performed between the left orbit and ear, and the skull was exposed. A craniotomy was performed to approach the left MCA. The stem of the MCA was irreversibly occluded by microbipolar electrocoagulation. Arterial pH, pCO_2, pO_2 (Corning 178, Corning, Gießen, Germany), mean arterial blood pressure (MABP) and plasma glucose concentrations were monitored up to 1 h after administration of the drug. Rectal temperature was maintained and controlled as mentioned above. After 7 days, rats were anesthetized with chloral hydrate (400 mg/kg, i.p.) and decapitated. Brains were removed from the skull and frozen in isopentane (Fluka, Buchs, Switzerland) on dry ice. Transverse sections of 20 μm thickness were taken every 500 μm using a cresyl violet stain to differentiate between intact and damaged brain tissue. The infarct volume (mm^3) was calculated from the infarct area of each section and the distance between succeeding sections. The animals received a single intraperitoneal injection of clenbuterol (0.1, 0.5 and 1 mg/kg) 5 h prior to vessel occlusion. Selegiline was administered intraperitoneally at a dose of 15 mg/kg once 2 h prior to MCA occlusion and for 8 consecutive days before induction of ischemia.

Statistics

All values are means ± SD or ± SEM of *n* experiments. One-way analysis of variance (ANOVA) combined with Scheffé's test or Duncan's test were used for multiple comparisons. Homogeneity of errors was determined using Bartlett's test.

Results

Clenbuterol and Selegiline Induce NGF in Cultured Rat Hippocampal Cells

A two-site enzyme immunoassay was used to measure NGF protein level in the medium of rat hippocampal cells. The culture medium was assayed directly in the ELISA system. Exposure to clenbuterol (1–100 μM) significantly enhanced the content of NGF in the culture medium: 33.3 ± 3.4 pg/ml in control cultures to 147.5 ± 20.6 pg/ml in clenbuterol (100 μM)-treated cultures, means ± SD, $n = 5$, $p < 0.001$ (Fig. 1A). An increase in NGF concentration in the medium was observed also after exposure of the hippocampal cells to very low concentrations of selegiline (10 pM–1 nM) for 6 h (Fig. 2A).

Exposure to Clenbuterol and Selegiline Reduces Glutamate-Induced Neurotoxicity in Rat Hippocampal Cultures

Exposure of rat hippocampal neurons after 14 days in vitro to 0.5 or 1 mM L-glutamate for 1 h in serum-free medium induced pronounced neurotoxicity evaluated morpho-

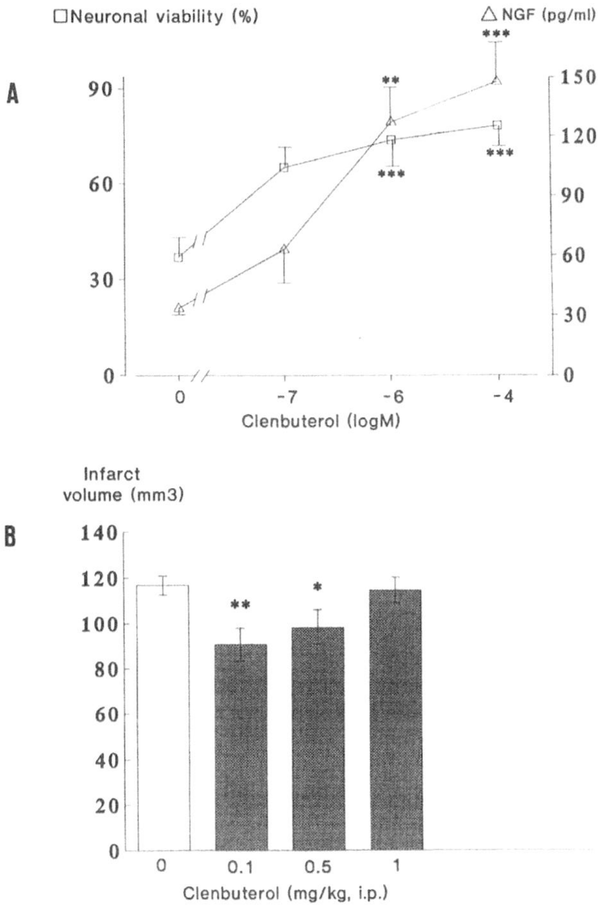

Fig. 1A,B. Clenbuterol causes neuroprotection in vitro and in vivo by induction of nerve growth factor (NGF). **A** Clenbuterol enhances NGF content in medium from cultured rat hippocampal cells and attenuates glutamate-induced excitotoxic damage. Mixed neuronal–glial hippocampal cells cultured for 14 days were exposed to clenbuterol (0.1–100 μM) and NGF content was determined in the medium 4 h later by NGF-enzyme linked immunosorbent assay (ELISA). The values are means ± SD from between five and seven experiments. Different from control: $^{***}p < 0.001$ and $^{**}p < 0.01$ by Scheffé's test. The sister cultures were exposed to 1 mM L-glutamate for 1 h. Neuronal damage was determined by trypan blue exclusion 18 h later. Clenbuterol was placed in the medium 4 h before excitotoxic injury and until 18 h after, including the period of glutamate exposure. Different from glutamate-exposed cells: $^{***}p < 0.001$ by Scheffé's test. **B** Clenbuterol reduces the infarct volume in the rat model of focal cerebral ischemia. Permanent middle cerebral artery (MCA) occlusion was performed in male Long Evans rats. The animals (eight to nine per group) received a single intraperitoneal injection of clenbuterol (0.1, 0.5 and 1 mg/kg) 5 h prior to vessel occlusion. Control animals (n = 17) received the vehicle only. Data are means ± SD. Different from control: $^{*}p < 0.05$ and $^{**}p < 0.01$ by Duncan's test

logically and by trypan blue staining 18 h later. In glutamate-exposed cultures neurites were decreased in number and some neurons had disintegrated into debris. Approximately 70% of neurons lost their membrane integrity and became stained with trypan blue 18 h after exposure to glutamate. Clenbuterol (1–100 μM), present in the culture

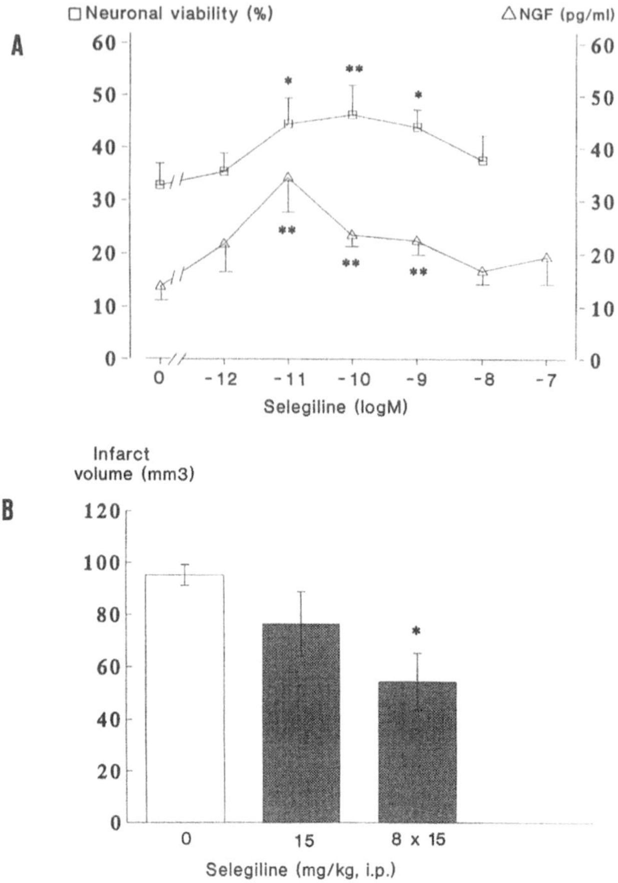

Fig. 2A,B. Selegiline induces nerve growth factor (NGF) and attenuates excitotoxic-ischemic neuronal damage. **A** Selegiline enhances NGF content in medium from rat hippocampal cells and protects hippocampal neurons against excitotoxic injury. Mixed neuronal-glial hippocampal cultures were grown for 14 days and then treated with various concentrations of selegiline for 6h in serum-free medium. NGF content was determined directly in the culture medium by NGF-enzyme linked immunosorbent assay (ELISA). Values are means ± SD from five different experiments. Different from controls: $^{**}p < 0.01$ by Scheffé's test. The sister cultures were exposed to 0.5 m*M* L-glutamate for 1h in serum-free medium. Neuronal damage was determined 18h later by trypan blue exclusion. Selegiline was present in the culture medium 6h before and up to 18h after excitotoxic injury, including the period of exposure to glutamate. Control cultures received the vehicle only. The values are means ± SD from five different experiments. Different from glutamate-exposed cultures: $^{*}p < 0.05$ and $^{**}p < 0.01$ by Scheffé's test. **B** Selegiline reduces the infarct volume in the rat model of focal cerebral ischemia. Permanent middle cerebral artery (MCA) occlusion was performed in male Long Evans rats. The animals (nine per group) were treated intraperitoneally with selegiline (15 mg/kg) once for 2h and for 8 consecutive days prior to occlusion of MCA. Control animals received the vehicle only. Data are means ± SEM. Different from control: $^{*}p < 0.05$ by Duncan's test

medium 4 h before and up to 18 h after induction of injury, greatly protected neurons against excitotoxic damage (Fig. 1A). The most pronounced protection was observed with 100 μM clenbuterol (15.4% ± 6.5% damaged neurons versus 63.3% ± 6.0% damaged neurons in glutamate-exposed controls, $n = 5$, $p < 0.001$ by Scheffé's test) (Fig. 1A). Treatment of the hippocampal cells with selegiline (10 pM–1 nM) for 6 h before and up to 18 h after the induction of excitotoxic injury also reduced significantly the number of damaged neurons (Fig. 2A).

Neuroprotective Activity of Clenbuterol and Selegiline in a Rat Model of Focal Cerebral Ischemia

Permanent occlusion of the MCA in rats induced ischemic cortical injury evaluated 2 days later as infarct volume (mm^3). A single intraperitoneal injection of clenbuterol (0.1 and 0.5 mg/kg) 5 h prior to occlusion of MCA significantly reduced the cortical infarct volume (Fig. 1B). Treatment with selegiline (15 mg/kg, i.p.) for 8 consecutive days before induction of focal cerebral ischemia also significantly reduced cortical ischemic injury of the rat brain (Fig. 2B).

Discussion

The synthesis of NGF in the brain is upregulated after different kinds of damage [11, 16, 22, 23], suggesting that this may be a protective mechanism to maintain neuronal survival. By pharmacological intervention the upregulation of NGF after central nervous system (CNS) damage might be further elevated or sustained. The adrenergic system appears to be involved in the regulation of NGF levels in the brain (see [26] for review). It has been shown that β-adrenergic stimulation increased the NGF mRNA and NGF protein in the medium of cultured astroglial cells and primary astrocytes via a cyclic adenosine monophosphate (cAMP)-dependent mechanism [10, 31, 33, 34, 44]. In vivo, a single intraperitoneal injection of clenbuterol elicited a two- to three-fold increase in NGF mRNA and NGF protein in rat cerebral cortex [12, 18]. These findings led to the speculation that β-adrenergic receptor stimulation could modulate the availability of NGF in the CNS. Therefore, we investigated whether clenbuterol can induce NGF synthesis in cultured rat hippocampal cells. Unstimulated hippocampal cells released detectable amounts of NGF into the medium, which were increased significantly (four- to five-fold) after incubation for 4 h with different concentrations of clenbuterol (1–100 μM), suggesting stimulation of NGF synthesis. Since astrocytes express β-adrenergic receptors [24] and respond to β-adrenergic stimulation with enhanced NGF synthesis and secretion, we speculate that the increased NGF level measured in medium of the mixed neuronal–glial hippocampal cultures after treatment with clenbuterol could come from activated glial cells.

We have shown previously that selegiline, an irreversible MAO-B inhibitor, at very low concentrations is able to induce NGF synthesis in cultured rat cortical astrocytes [36]. Incubation of hippocampal cells for 6 h with the same very low concentrations of selegiline (10 pM–1 nM) also led to enhanced content of NGF in the culture medium as determined by NGF-ELISA. This effect of selegiline did not seem to depend on the inhibition of MAO-B. It has been shown previously that selegiline alters gene expression and protein synthesis in astrocytes [19, 37, 42] at concentrations insufficient to cause inhibition of MAO-B. Importantly, it has been demonstrated that this drug induces the process of reactive astrogliosis after mechanical lesions both in vitro [37] and in vivo [4] and increases the expression of ciliary neurotrophic factor (CNTF) [37] in cultured astrocytes. Activated astrocytes produce several growth factors that support neuronal survival and NGF is only one of them. Therefore, incubation of hippocampal cells with selegiline might cause an increased synthesis and release of NGF from activated astrocytes.

Furthermore, we attempted to determine whether clenbuterol and selegiline could protect hippocampal neurons against glutamate-induced excitotoxic injury by increased NGF synthesis. It has been demonstrated that NGF itself attenuates hypoglycemic–excitotoxic and iron-induced degeneration in cultured hippocampal cells [6, 7, 25, 45]. Clenbuterol also attenuated glutamate-induced excitotoxicity in vitro as has also been previously shown [35]. When this drug (1–100 μM) was present in the medium 4 h before until 18 h after induction of injury, the hippocampal neurons were greatly protected from glutamate-induced degeneration. The neuroprotective activity of clenbuterol was abolished by propranolol (data not shown), suggesting that the induction of NGF and neuroprotection caused by this drug are mediated via β-adrenergic receptor activation. Selegiline was also able to protect hippocampal neurons from excitotoxic damage. The neuroprotective activity of selegiline was demonstrated at the same concentrations (10 pM–1 nM) which increase NGF content in the culture medium, suggesting that NGF might underlie the neuroprotective action of this compound. It has been shown that selegiline exerts some neurorescuing and "trophic-like" activities which do not depend upon its inhibition of MAO-B [1, 32]. At very low concentrations selegiline induces reactive gliosis and expression of CNTF in cultured astrocytes. Therefore, the pharmacological basis for MAO-B-independent activities of selegiline could be associated with enhanced synthesis and release of neurotrophic factors from activated astrocytes.

The neuroprotective activity of clenbuterol and selegiline was also demonstrated in vivo in a rat model of focal cerebral ischemia. Pretreatment with clenbuterol (0.1 and 0.5 mg/kg) for 5 h prior to occlusion of the MCA was enough to protect rat cerebral cortex from injury due to the ischemic insult. Since it has been reported that clenbuterol increased the level of NGF in rat cerebral cortex [12, 18], we speculate that the observed neuroprotection is due to enhanced amounts of NGF in this brain region. Other in vivo studies have indicated that NGF can reduce neuronal damage in animal models of ischemic brain injury [5, 13, 29]. Neuroprotection against MCA occlusion-induced ischemic damage was also observed after intraperitoneal administration of selegiline (8×15 mg/kg) prior to induction of damage. It has been reported that selegiline induces reactive astrogliosis that occurs following mechanical damage to brain tissue [3, 4, 19]. Activated astrocytes produce and release a number of growth factors [2, 14] that support neuronal survival after damage. This "astrocyte-kinetic" effect of selegiline may underlie it's neuroprotective and "trophic-like" activities in CNS independent of MAO-B inhibition. Since selegiline was able to increase NGF synthesis in cultured rat hippocampal cells and to protect them against excitotoxic damage we suggest that induction of NGF synthesis in vivo may contribute to the neuroprotective activity of this drug.

In summary, the present study indicates that two lipophilic drugs, clenbuterol, a $\beta 2$-adrenergic receptor agonist, and selegiline, a MAO-B inhibitor, enhanced the concentration of NGF in the medium of rat hippocampal cells and protected cultured hippocampal neurons against glutamate-induced excitotoxic damage. In vivo these compounds protected rat cerebral cortex from ischemic insult due to permanent MCA occlusion. NGF could mediate the neuroprotective activity demonstrated by clenbuterol and selegiline both in vitro and in vivo. Since NGF cannot cross the BBB, stimulating its endogenous synthesis by lipophilic drugs could become an alternative therapy for stroke and trauma-induced neuronal damage as well as for chronic neurodegenerative diseases.

Acknowledgment. This work was supported by the Deutsche Forschungsgemeinschaft (Bonn, Germany), grant Kr. 354/16-1.

References

1. Ansari KS, Yu PH, Kruck TPA, Tatton WG (1993) Rescue of axotomized immature rat-facial motoneurons by R(–)-deprenyl: stereospecificity and independence from monoamine oxidase inhibition. J Neurosci 13: 1043–1052
2. Bakhit C, Armanini M, Bennett GL, Wong WLT, Hansen SE, Taylor R (1991) Increase in glia-derived nerve growth factor following destruction of hippocampal neurons. Brain Res 560: 76–83
3. Biagini G, Zoli M, Fuxe K, Agnati LF (1993) l-Deprenyl increases GFAP immunoreactivity selectively in activated astrocytes in rat brain. Neuroreport 4: 955–958
4. Biagini G, Frasoldati A, Fuxe K, Agnati LF (1994) The concept of astrocyte-kinetic drug in the treatment of neurodegenerative diseases: evidence for l-deprenyl-induced activation of reactive astrocytes. Neurochem Int 25: 17–22
5. Buchan AM, Williams L, Bruederlin B (1990) Nerve growth factor: pretreatment ameliorates ischemic hippocampal neuronal injury. Stroke 21: 177
6. Cheng B, Mattson MP (1991) NGF and bFGF protect rat hippocampal and human cortical neurons against hypoglycemic damage by stabilizing calcium homeostasis. Neuron 7: 1031–1041
7. Cheng B, Mattson MP (1992) Glucose deprivation elicits neurofibrillary tangle-like antigenic changes in hippocampal neurons: prevention by NGF and bFGF. Exp Neurol 117: 114–123
8. Choi DW (1987) Ionic dependence of glutamate neurotoxicity. J Neurosci 7: 369–379
9. Choi DW (1988) Calcium-mediated neurotoxicity: relationship to specific channel types and role in ischemic damage. Trends Neurosci 11: 465–469
10. Dal Toso R, De Bernardi MA, Brooker G, Costa E, Mocchetti I (1988) Beta-adrenergic and prostaglandin receptor activation increases nerve growth factor mRNA content in C6-2B rat astrocytoma cells. J Pharmacol Exp Ther 246: 1190–1193
11. DeKosky ST, Goss JR, Miller PD, Styren SD, Kochanek PM, Marion D (1994) Upregulation of nerve growth factor following cortical trauma. Exp Neurol 130: 173–177
12. Follesa P, Mocchetti I (1992) Regulation of basic fibroblast growth factor and nerve growth factor mRNA by β-adrenergic receptor activation and adrenal steroids in rat central nervous system. Mol Pharmacol 43: 132–138
13. Frim DM, Short MP, Rosenberg WS, Simpson J, Breakfield XO, Isacson O (1993) Local protective effect of nerve growth factor-secreting fibroblasts against excitotoxic lesions in the rat striatum. J Neurosurg 78: 267–273
14. Giulian D, Vaca K, Corpuz M (1993) Brain glia release factors with opposing actions upon neuronal survival. J Neurosci 13: 29–37
15. Greenamyre JT, Young AB (1989) Excitatory amino acids and Alzheimer's disease. Neurobiol Aging 10: 593–602

16. Gwag BJ, Sessler F, Kimmerer K, Springer JE (1994) Neurotrophic factor mRNA expression in dentate gyrus is increased following angular bundle transection. Brain Res 647: 23–29
17. Hartikka J, Hefti F (1988) Development of septal cholinergic neurons in culture: plating density and glial cells modulate effects of NGF on survival, fiber growth, and expression of transmitter-specific enzymes. J Neurosci 8: 2967–2985
18. Hayes VY, Isackson PJ, Fabrazzo M, Follesa P, Mocchetti I (1995) Induction of nerve growth factor and basic fibroblast growth factor mRNA following clenbuterol: contrasting anatomical and cellular localization. Exp Neurol 132: 33–41
19. Ju WYH, Holland DP, Tatton WG (1994) (–)-Deprenyl alters the time course of death of axotomized facial motoneurons and the hypertrophy of neighboring astrocytes in immature rats. Exp Neurol 126: 233–246
20. Koh JI, Choi DW (1988) Vulnerability of cultured cortical neurons to damage by excitotoxins: Differential susceptibility of neurons containing NADPH-diaphorase. J Neurosci 8: 2153–2163
21. Korsching S, Thoenen H (1983) Nerve growth factor in sympathetic ganglia and corresponding target organs of the rat: correlation with density of sympathetic innervation. Proc Natl Acad Sci USA 80: 3513–3516
22. Lauterborn JC, Isackson PJ, Gall CM (1994) Seizure-induced increases in NGF mRNA exhibit different time courses across forebrain regions and are biphasic in hippocampus. Exp Neurol 125: 22–40
23. Lorez H, Keller F, Ruess G, Otten U (1989) Nerve growth factor increases in adult rat brain after hypoxic injury. Neurosci Lett 98: 339–344
24. Mantyh PW, Rogers SD, Allen CJ, Catton MD, Ghilard JR, Levin LA, Maggio JE, Vigna SR (1995) β_2-adrenergic receptors are expressed by glia in vivo in the normal and injured central nervous system in the rat, rabbit, and human. J Neurosci 15: 152–164
25. Mattson MP, Lovell MA, Furukawa K, Markesbery WR (1995) Neurotrophic factors attenuate glutamate-induced accumulation of peroxides, elevation of intracellular Ca^{2+} concentration, and neurotoxicity and increase antioxidant enzyme activities in hippocampal neurons. J Neurochem 65: 1740–1751
26. Mocchetti I (1991) Theoretical basis for a pharmacology of nerve growth biosynthesis. Ann Rev Pharmacol Toxicol 32: 303–328
27. Olney JW, Zorumski CF, Price MT, Wang GG, Labruyere J (1990) Excitotoxicity of l-DOPA and 6-OH-DOPA: Implications for Parkinson's and Huntington's diseases. Exp Neurol 108: 269–272
28. Ossowska K (1994) The role of excitatory amino acids in experimental models of Parkinson's disease. J Neural Transm [P-D Sect] 8: 39–71
29. Pechan PA, Yoshida T, Panahian N, Moskowitz MA, Breakefield XO (1995) Genetically modified fibroblasts producing NGF protect hippocampal neurons after ischemia in the rat. Neuroreport 6: 669–672
30. Rothstein JD, Martin LJ, Kunel RW (1992) Decreased glutamate transport by the brain and spinal cord in amyotrophic lateral sclerosis. N Engl J Med 326: 1416–1468
31. Rudge JS, Morrissey D, Lindsay RM, Pasnikowski EM (1994) Regulation of ciliary neurotrophic factor in cultured rat hippocampal astrocytes. Eur J Neurosci 6: 218–229
32. Salo PT, Tatton WG (1992) Deprenyl reduces the death of motoneurons caused by axotomy. J Neurosci Res 31: 394–400
33. Schwartz JP, Costa E (1977) Regulation of nerve growth factor content in C6 glioma cells by β-adrenergic receptor stimulation. Naunyn Schmiedebergs Arch Pharmacol 300: 123–129
34. Schwartz JP, Mishler K (1990) Beta-adrenergic receptor regulation, through cyclic AMP, of nerve growth factor expression in rat cortical and cerebellar astrocytes. Cell Mol Neurobiol 10: 447–457
35. Semkova I, Rami A, Krieglstein J (1994) Neuroprotective activity of clenbuterol against excitotoxic neuronal damage. In: Krieglstein J, Oberpichler-Schwenk H (eds) Pharmacology of cerebral ischemia. Medpharm Scientific, Stuttgart, pp 397–406
36. Semkova I, Krieglstein J (1995) Induction of NGF in rat cortical astrocytes by isoproterenol, clenbuterol and selegiline. J Cereb Blood Flow Metab 15 [Suppl 1], Abstr 573
37. Seniuk NA, Henderson JT, Tatton WG, Roder JC (1994) Increased CNTF gene expression in process-bearing astrocytes following injury is augmented by R(–)-deprenyl. J Neurosci Res 37: 278–286
38. Shelton DL, Reichardt LF (1986) Studies on the expression of the β-nerve growth factor (NGF) gene in the central nervous system: level and regional distribution of NGF mRNA suggests that

NGF functions as a trophic factor for several distinct populations of neurons. Proc Natl Acad Sci USA 83: 2714–2718
39. Shigeno T, Mima T, Takaura K, Graham DI, Kato G, Hashimoto Y, Furukawa S (1991) Amelioration of delayed neuronal death in the hippocampus by nerve growth factor. J Neurosci 11: 2914–2919
40. Simon RP, Swans JH, Griffith T, Meldrum BS (1984) Blockade of N-methyl-D-aspartate receptors may protect against ischemic damage in the brain. Science 226: 850–852
41. Tamura A, Graham DJ, McCulloch J, Teasdale GM (1981) Focal cerebral ischemia in the rat: regional cerebral blood flow determined by [^{14}C]-iodoantipyrine autoradiography following middle cerebral artery occlusion. J Cereb Blood Flow Metab 1: 61–69
42. Tatton WG, Ju WYL, Holland DP, Tai C, Kwan M (1994) (–)-Deprenyl reduces PC12 cell apoptosis by inducing new protein synthesis. J Neurochem 63: 1572–1575
43. Whittemore SR, Seiger A (1987) The expression, localization and functional significance of β-nerve growth factor in the central nervous system. Brain Res Rev 12: 439–446
44. Zafra F, Lindholm D, Castren E, Hartikka J, Thoenen H (1992) Regulation of brain derived neurotrophic factor and nerve growth factor mRNA in primary cultures of hippocampal neurons and astrocytes. J Neurosci 12: 4793–4799
45. Zhang Y, Tatsuno T, Carney JM, Mattson MP (1993) Basic FGF, NGF, and IGFs protect hippocampal and cortical neurons against iron-induced degeneration. J Cereb Blood Flow Metab 13: 378–388

Stress Response and Ischemic Tolerance in the Brain

M. Matsumoto

Summary

In order to elucidate the cellular stress responses of the brain induced by ischemia-reperfusion, we investigated the induction of stress proteins in the cultured astrocytes exposed to hypoxia-reoxygenation (H-R). In a separate in vivo study, we examined the inducibility of ischemic tolerance and heat shock protein 72 (HSP72) in moderately symptomatic gerbils by unilateral preconditioning ischemia. In the former studies, we first attempted to clarify the role of the most abundant stress protein, glucose-regulated protein (GRP) 78, in H-R. Introduction of antisense oligonucleotide to GRP78 into astrocytes suppressed H-R-induced astrocyte release of interleukin 6 (IL-6), and cell lysates from H-R astrocytes displayed induction of IL-6, which was present in the samples immunoprecipitated with anti-GRP78, suggesting the role of GRP78 as a molecular chaperone for sustaining generation of IL-6 under the stress of H-R. Another unique stress protein, oxygen-regulated protein 150 (ORP150), was purified and through several experimental steps we cloned the ORP150 gene successfully. These stress proteins may contribute to the adaptive response of astrocytes to ischemic stress, thereby ultimately contributing to the enhanced survival of neurons. In the latter studies, ischemic tolerance could be induced in the ipsilateral caudate-putamen of gerbils with moderate neurological signs secondary to preconditioning unilateral ischemia which was reversible but strong enough to induce HSP72. The present investigation strongly suggested an important role of stress proteins in the cellular stress responses of the brain subjected to ischemia-reperfusion.

Introduction

Since the first documentation of the induction of heat shock protein (HSP) following transient cerebral ischemia [18], HSP induction in experimental brain ischemia has been observed not only in the vulnerable neuronal cells but also in non-neuronal cells such as astrocytes and endothelial cells [4, 20], depending on the degree of ischemic stress. These findings suggest that cellular elements in the central nervous system respond to an altered environment in a manner analogous to that of cultured cells. In this report, our research is focused initially on an important aspect of hypoxia-

Division of Stroke Research, First Department of Medicine and Department of Neurology, Osaka University Medical School, 2-2 Yamada-oka, Suita, Osaka 565, Japan

Maturation Phenomenon in Cerebral Ischemia II
U. Ito et al. (Eds.)

induced redirection of cellular biosynthetic mechanisms through studies of the role of glucose-regulated protein 78 (GRP78) as a molecular chaperone for release of interleukin 6 (IL-6) by astrocytes, and on the purification and characterization of a novel stress protein, the 150-kDa oxygen-regulated protein (ORP150). Also, special attention has been paid to the relationship between ischemic tolerance and heat shock protein 72 (HSP72) through the development of a double ischemia model using gerbils with unilateral preconditioning ischemia.

Stress Proteins in Cultured Astrocytes

As the most abundant cell type in the central nervous system, astrocytes are positioned to nurture and protect neurons in response to several stresses, including ischemia. These considerations have led us to investigate stress responses in the cultured astrocytes exposed to hypoxia–reoxygenation (H–R). In our previous studies, we observed the elaboration of the neuroprotective cytokine interleukin 6 (IL-6) [13], and the induction of several stress proteins (HSPs, GRPs, ORPs, etc.) [2] in cultured astrocytes exposed to H–R. Among several stress proteins induced by H–R, 78-kDa protein is the most abundant and has been identified as GRP78 through several lines of evidence, including parallel induction by 2-deoxyglucose and a calcium ionophore, microprotein sequence, western blotting and suppression by antisense oligonucleotide application. Our data also suggested the role of this protein as a molecular chaperone which facilitates the elaboration of IL-6 by reoxygenated astrocytes.

Rat astrocytes were obtained from neonatal rat brain as described previously [13]. After the removal of microglia cells and type II astrocytes, type I astrocytes (>98% based on immunohistochemical detection of glial fibrillary acidic protein) were used for the following experiments. Sodium dodecyl sulphate-polyacrylamide gel electrophoresis (SDS-PAGE) of metabolically labeled astrocytes exposed to H–R demonstrated five induced–enhanced bands within 12h, corresponding to molecular weights of 28kDa, 33kDa, 78kDa, 94kDa and 150kDa (Figs. 1, 2). Addition of 2-deoxyglucose to the medium of normoxic astrocytes also demonstrated prominent induction of the 78-kDa protein (Figs. 1, 2), suggesting that its expression might be related to glucose metabolism. After several purification steps, this protein was subjected to N-terminal sequencing and the sequence (Glu-Glu-Glu-Asp-Lys-Lys-Glu-Asp-Val-Gly) obtained was the same as that of the 78-kDa rat GRP78 [17]. Incubation of astrocytes with antisense oligonucleotide for GRP78 largely prevented enhanced expression of this protein in H–R. GRP78 has been known to work as an intracellular molecular chaperone in the endoplasmic reticulum and to bind secretory proteins, promoting their proper folding and translocation and then dissociating when ATP levels are restored [1, 11]. This led us to speculate that GRP78 might function as a molecular chaperone in hypoxic astrocytes by promoting stabilization of molecules such as the neuroprotective cytokine IL-6, whose expression is enhanced in astrocytes subjected to H–R [13]. Therefore, we investigated the effect of antisense oligonucleotide for GRP78 on expression of IL-6 by astrocytes exposed to H–R [3]. Addition of GRP78 antisense oligonucleotide at the beginning of hypoxia significantly suppressed the expression of IL-6 in culture media of astrocytes exposed to H–R, while no

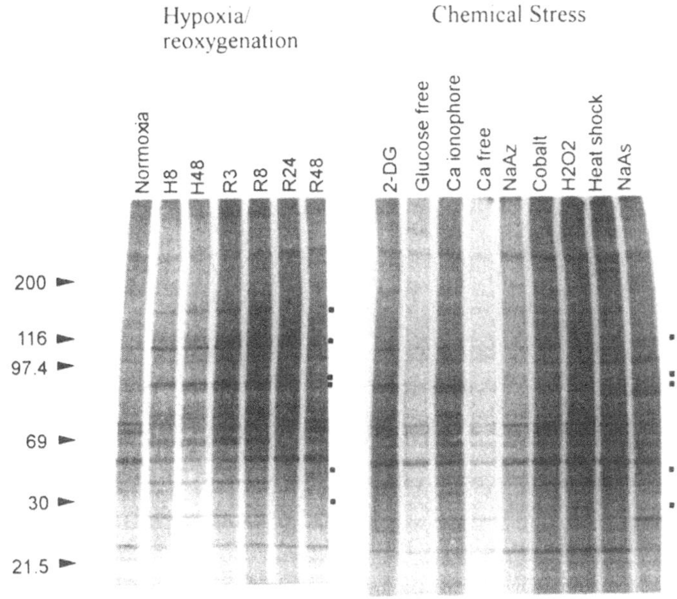

Fig. 1. Analysis of metabolically labeled cultured astrocytes exposed to hypoxia-reoxygenation (H–R) (*left panel*) and chemical stresses (*right panel*). *Left panel*, astrocytes cultured in methionine-poor medium were either exposed to hypoxia alone for 8 h or 48 h (*H8* or *H48*, respectively), or exposed to hypoxia for 48 h and then reoxygenated for 3, 8, 24 or 48 h (*R3*, *R8*, *R24* or *R48*, respectively). Metabolic labeling was accompanied by addition of [^{35}S]methionine (200 μCi/ml) 3 h before the end of the experiment. Cells were solubilized in a reduced sodium dodecyl sulphate (SDS) sample buffer, subjected to SDS-polyacrylamide gel electrophoresis (PAGE) (5%–20%; about 1 μg protein/lane), followed by autoradiography. The *arrowheads* at the far left side of the gel indicate the migration of simultaneously-run molecular weight standards: trypsin inhibitor (21.5 kDa), carbonic anhydrase (30 kDa), ovalbumin (46 kDa), bovine serum albumin (69 kDa), phosphorylase b (97.4 kDa), and myosin (200 kDa). *Left panel*, normoxic astrocytes were incubated in the presence of 2-deoxyglucose (20 m*M*; *2-DG*), glucose-free medium (*Glucose free*), the calcium ionophore A23178 (10 μM; *Ca ionophore*), calcium-free medium (*Ca free*), sodium azide (1 m*M*; *NaAz*), and Co C12 (0.1 m*M*; *Cobalt*) for 12 h at 37°C. In these experiments, [^{35}S]methionine (200 μCi/ml) was added to each culture for 3 h befroe harvest. Other studies employed astrocytes exposed to either hydrogen peroxide (50 μM; *H2O2*) for 60 min, sodium arsenite (0.1 m*M*; *NaAs*) for 3 h, or elevated temperature (43°C; *Heat shock*) for 3 h. The incubation was continued for an additional 3 h at 37°C in the presence of [^{35}S]methionine to allow for labeling, and samples were processed as above for SDS-PAGE and autoradiography. The migration of bands with M_r's corresponding to 150, 94, 78, 72, 33 and 28 kDa, whose expression was induced by H–R or chemical stresses is indicated by the *dots*

significant effect was observed by adding the same concentration of sense GRP78. We further analyzed lysates from astrocytes subjected to H–R by gel filtration and could detect IL-6 activity in the fractions containing larger molecular size proteins (fraction 2–6) when treated with ATP, though no IL-6 activity was observed before ATP treatment. Formation of a complex between IL-6 and GRP78 has been further confirmed by demonstrating IL-6 activity after addition of ATP to the samples pooled from gel filtration fractions (2–6) of lysates from H–R astrocytes and immunoprecipitated with anti-GRP78. These results strongly suggested a crucial role of GRP78 as a molecular

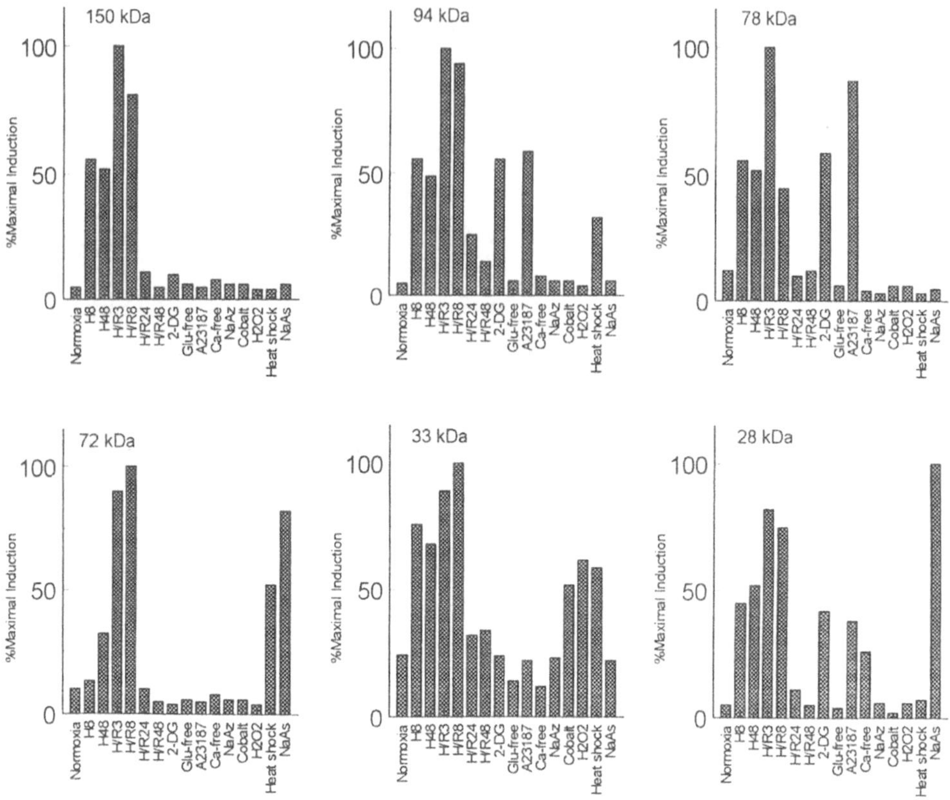

Fig. 2. Densitometric analysis of autoradiograms from metabolically labeled astrocytes exposed to hypoxia-reoxygenation (H–R) or chemical stresses: effect on expression of M_r 28-, 33-, 72-, 78-, 94- and 150-kDa stress proteins. Astrocytes were subjected to H–R or chemical stresses as described in Fig. 1, metabolically labeled with [^{35}S]methionine, subjected to sodium dodecyl sulphate-polyacrylamide gel electrophoresis (SDS-PAGE), and autoradiography. Autoradiograms were analyzed by laser densitometry and the density of each stress protein-band was semiquantitatively assessed by Quality-One. Values are expressed as a percent of maximal induction observed within the same experimental series. The mean of two different autoradiograms is shown, and the experiment was repeated three times

chaperone for preventing denaturation and aggregation of secretory proteins such as IL-6 during the cellular response to environmental stress such as H–R.

The above studies on GRP78 emphasized the potentially important role of stress proteins in the cellular adaptation to several environmental perturbations and led us to characterize other stress proteins induced in the astrocytes exposed to H–R. Among these stress proteins ORP150 seems to be quite unique [10]. This stress protein was induced in the astrocytes only by oxygen deprivation stress and not by a range of other stimuli, including heat shock, hydrogen peroxide, heavy metals and 2-deoxyglucose (Figs. 1, 2). For purification of ORP150, we used lysates from cultured astrocytes exposed to hypoxia for 48 h as starting material. Fractions eluted from fast protein liquid chromatography (FPLC) Mono Q that contained ORP150 were pooled and concentrated by ultrafiltration and about 200 μg of protein was applied to preparative SDS-PAGE. N-terminal sequencing of the band with a molecular mass corresponding to ORP150 was performed using an automated peptide sequencing system. A search of the SWISSPROT database indicated that the N-terminal amino acid sequence of ORP150 (Leu-Ala-Val-Met-Ser-Val-Asp-Leu-Gly-Ser-Glu-Ser-Met-Lys-Val) was novel. We then made the degenerated polymerase chain reaction (PCR) primer and by screening the cDNA library of hypoxic astrocytes with this PCR product we were able to successfully clone the ORP150 gene (GenBank No. U41853). This protein has adenosine triphosphatase (ATPase) domains and an evoked response (ER)-retension signal in its primary structure deduced from cDNA sequence. To further analyze expression of ORP150, polyclonal antibodies were raised in rabbits. Immunohistochemical study using the specific antibody showed localization of this stress protein in the endoplasmic reticulum, and the time course of expression of ORP150 in astrocytes subjected to H–R paralleled that of GRP78. We also confirmed the expression of this protein in the ischemic mouse brain subjected to middle cerebral artery (MCA) occlusion for 24 h. The pathophysiological roles of this unique stress protein in cerebral ischemia should be investigated further in the near future.

Ischemic Tolerance and Stress Proteins in Unilateral Ischemia

Since the first documentation of the "ischemic tolerance" phenomenon by Kitagawa et al. [7] and the subsequent demonstration of the induction of HSP72 in hippocampal CA1 neurons by preconditioning reversible ischemia [5], the relationship between this unique cell defense phenomenon and the induction of HSPs has been vigorously investigated, mainly by using transient forebrain ischemia as a preconditioning ischemia. In our previous study [19], we demonstrated the induction of ischemic tolerance by prolonged but mild hypoperfusion in asymptomatic gerbils with unilateral carotid occlusion. In the current study [9], we have explored the presence of ischemic tolerance in the cerebral hemisphere after unilateral carotid occlusion in moderately symptomatic gerbils, which should moderately reduce focal cerebral blood flow in the affected hemisphere [14]. We first examined the extent of ischemic lesions neuropathologically in gerbils with moderate stroke symptoms following 7 days of recirculation after unilateral carotid occlusion for 10 min and 30 min ($n = 6$ and 12, respectively). Both conventional methods (hematoxylin-eosin and cresyl violet staining) and immunohistochemical staining for microtubule-associated protein 2 [6] were employed for the detection of ischemic lesions. Although both 10 min and 30 min of ischemia showed ischemic lesions in the CA1 and/or CA4 subfields of the affected hippocampus of all gerbils with moderate neurological signs, ischemic lesions in the caudate-putamen were detected in 11 of 12 moderately symptomatic gerbils with 30-min unilateral ischemia, and not in those with 10-min ischemia. We also examined the induction of HSP72 in the affected hemisphere of the moderately symptomatic gerbils at 2 days after transient unilateral carotid occlusion for 10 min, and HSP72 induction was clearly demonstrated in the neurons of the caudate-putamen and cortex. Based on these results, we investigated the induction of ischemic tolerance phenomenon in a double unilateral ischemia model in gerbils with moder-

Table 1. Grading of ischemic severity based on the expression of stress gene and tolerance (from [8])

Grade 0	Subtle and non-stressful	Neurons survive without showing any stress response
Grade 1	Reversible but strong enough to cause stress reaction	Neurons survive but show stress response and acquire tolerance against a subsequent ischemic stress
Grade 2	Irreversible but prevented by the "grade 1" ischemic pretreatment	Neurons die unless pretreated by the "grade 1" ischemic stress, but show gene expression during their process leading to death
Grade 3	Irreversible and too strong to be protected by the "grade 1" ischemic pretreatment	Neurons die while showing gene transcription with or without producing the translation product
Grade 4	Irreversible and acutely necrotic	Neurons die without showing any response

ate neurological signs (30-min unilateral carotid occlusion 2 days after 10-min preconditioning ischemia). The mean percent area of ischemic lesions in the caudate-putamen as detected by loss of immunoreaction for microtubule-associated protein (MAP2) was 9.8% in gerbils with preconditioning ischemia, which was a significantly smaller loss than that in gerbils without preconditioning ischemia (68.6%). The results clearly show that ischemic tolerance can be induced in the caudate-putamen of gerbils with moderate stroke symptoms, where 30-min ischemia consistently caused neuronal death, but that it can be prevented by 10-min preconditioning ischemia which is reversible but strong enough to induce HSP72. As described by Kitagawa et al. [8], ischemic stress can be categorized into five different grades depending on the extent of induction of HSP72 gene expression and tolerance (Table 1). However, there are several candidates other than HSP gene expression for induction of ischemic tolerance. Neurotrophic factors such as NGF, brain-derived neurotrophic factor (BDNF), neurotropin (NT-3), etc., neurotrophic cytokines including IL-6, radical scavenging enzymes including Mn-superoxide dismutase (SOD) and products of several specific genes (c-jun, bcl-2, etc.), which are known to be induced by ischemia–reperfusion [12, 15, 16, 21, 22], may also be involved in the acquisition of ischemic tolerance.

References

1. Dorner AJ, Wasley LC, Kaufman RJ (1990) Protein dissociation of GRP78 and secretion are blocked by depletion of cellular ATP levels. Proc Natl Acad Sci USA 87: 7429–7432
2. Hori O, Matsumoto M, Maeda Y, Ueda H, Ohtsuki T, Stern DM, Kinoshita T, Ogawa S, Kamada T (1994) Metabolic and biosynthetic alterations in cultured astrocytes exposed to hypoxia/reoxygenation. J Neurochem 62: 1489–1495
3. Hori O, Matsumoto M, Kuwabara K, Maeda Y, Ueda H, Ohtsuki T, Kinoshita T, Ogawa S, Stern DM, Kamada T (1996) Exposure of astrocytes to hypoxia/reoxygenation enhances expression of glucose-related protein 78 facilitating astrocyte release of the neuroprotective cytokine interleukin 6. J Neurochem 66: 973–979

4. Kinouchi H, Sharp FR, Hill MP, Koistinaho J, Sagar SM, Chan PH (1993) Induction of 70 kDa heat shock protein and hsp70 mRNA following transient focal cerebral ischemia in the rat. J Cereb Blood Flow Metab 13: 105–115
5. Kirino T, Tsujita Y, Tamura A (1991) Induced tolerance to ischemia in gerbil hippocampal neurons. J Cereb Blood Flow Metab 11: 299–307
6. Kitagawa K, Matsumoto M, Niinobe M, Mikoshiba K, Hata R, Ueda H, Handa N, Fukunaga R, Isaka Y, Kimura K, Kamada T (1989) Microtubule-associated protein 2 as a sensitive marker for cerebral ischemic damage. – immunohistochemical investigation of dendritic damage. Neuroscience 31: 401–411
7. Kitagawa K, Matsumoto M, Tagaya M, Hata R, Ueda H, Niinobe M, Handa N, Fukunaga R, Mikoshiba K, Kamada T (1990) Ischemic tolerance phenomenon found in the brain. Brain Res 528: 21–24
8. Kiatagwa K, Matsumoto M, Kamada T, Yanagihara T (1996a) Ischemic vulnerability and tolerance in the brain. In: Kamada T, Shiga T, McCuskey RS (eds) Tissue perfusion and organ function: Ischemia/reperfusion injury. Elsevier Science, Amsterdam, pp 85–94
9. Kitagawa K, Matsumoto M, Matsushita K, Mandai K, Mabuchi T, Yanagihara T, Kamada T (1996b) Ischemic tolerance in moderately symptomatic gerbils after unilateral carotid occlusion. Brain Res 716: 39–46
10. Kuwabara K, Matsumoto M, Ikeda J, Hori O, Ogawa S, Maeda Y, Kitagawa K, Imuta N, Kinoshita T, Stern DM, Yanagi H, Kamada T (1996) Purification and characterization of a novel stress protein, the 150-kDa oxygen-regulated protein (ORP150), from cultured rat astrocytes and its expression in ischemic mouse brain. J Biol Chem 271: 5025–5032
11. Lee AS (1992) Mammalian stress response: induction of the glucose regulated protein family. Curr Opin Cell Biol 4: 267–273
12. Lindvall O, Ernfors P, Bengzon J, Kokaia Z, Smith ML, Siesjo BK, Persson H (1992) Differential regulation of mRNAs for nerve growth factor, brain-derived neurotrophic factor, and neurotrophin 3 in the adult rat brain following cerebral ischemia and hypoglycemic coma. Proc Natl Acad Sci USA 89: 648–652
13. Maeda Y, Matsumoto M, Ohtsuki T, Kuwabara K, Ogawa S, Hori O, Shui DY, Kinoshita T, Kamada T, Stern D (1994) Hypoxia-reoxygenation mediated induction of interleukin-6 in cultured rat astrocytes and expression in ischemic gerbil brain: a paracrine mechanism enhancing neuron survival. J Exp Med 180: 2297–2308
14. Matsumoto M, Hatakeyama T, Yamamoto K, Yanagihara T (1988) Prediction of stroke before and after unilateral occlusion of the common carotid artery in gerbils. Stroke 19: 490–497
15. Matsuyama T, Michishita H, Nakamura H, Tsuchiyama M, Shimizu S, Watanabe K, Sugita M (1993) Induction of copper-zinc superoxide dismutase in gerbil hippocampus after ischemia. J Cereb Blood Flow Metab 13: 135–144
16. Merlio JP, Ernfors P, Kokaia Z, Middlemas DS, Bengzon J, Kokaia M, Smith ML, Siesjo BK, Hunter T, Lindvall O, Persson H (1993) Increased production of the TrkB protein tyrosine kinase receptor after brain insults. Neuron 10: 151–164
17. Munro S, Pelham HRB (1986) An Hsp70 like protein in the ER: identity with the 78 kd glucose-regulated protein and immunoglobin heavy chain binding protein. Cell 46: 291–300
18. Nowak TS Jr (1985) Synthesis of a stress protein following transient ischemia in the gerbil. J Neurochem 45: 1635–1641
19. Ohtsuki T, Matsumoto M, Kitagawa K, Taguchi A, Maeda Y, Hata R, Ogawa S, Ueda H, Handa N, Kimura K, Kamada T (1993) Induced resistance and susceptibility to cerebral ischemia in gerbil hippocampal neurons by prolonged but mild hypoperfusion. Brain Res 614: 279–284
20. Sharp FR, Lowenstein D, Simon R, Hisanaga K (1991) Heat shock protein hsp72 induction in cortical and striatal astrocytes and neurons following infarction. J Cereb Blood Flow Metab 11: 621–627
21. Shimazaki K, Ishida A, Kawai N (1994) Increase in bcl-2 oncoprotein and the tolerance to ischemia-induced neuronal death in the gerbil hippocampus. Neurosci Res 20: 95–99
22. Sommer C, Gass P, Kiessling M (1995) Selective c-JUN expression in CA1 neurons of the gerbil hippocampus during and after acquisition of an ischemia-tolerant state. Brain Pathol 5: 135–144

Preconditioning Depresses Excitatory Cell Signaling Following the Second Ischemic Insult

M. SHAMLOO and T. WIELOCH

Summary

Ischemic tolerance was induced in the rat using the two-vessel occlusion model of cerebral ischemia. Three minutes of preconditioning ischemia with 2–3 days of interischemic interval protected the brain from a subsequent period of lethal ischemia of 10–15 min duration. Calcium-calmodulin protein kinase II (CaMKII) is translocated from its cytosolic form to the membrane-bound form during ischemia. In the CA1 region the translocated protein is bound to the membrane fraction for up to 24 h during recovery, without replenishment in the cytosolic fraction. Preconditioning ischemia decreases and normalizes the membrane bound CaMKII levels after 24 h of recovery. Tyrosine phosphorylation of the NMDA receptor (NR2A/B) subunit is enhanced during recovery following ischemia. In preconditioned brain the tyrosine phosphorylation is decreased and normalized by 24 h of recovery. The NR2A/B subunit levels are markedly diminished after 24 h of recovery. We propose that during the preconditioning, protective factors (growth factors and cytokines) are induced in neurons and/or glia cells, which in turn stimulate intracellular neuroprotective mechanisms such as enhancers of proteins synthesis or bcl-2 expression, which in concert with processes activated by the second ischemic insult downregulate deleterious processes including those elevating intracellular calcium ion concentrations.

Introduction

Both intracellular and extracellular signaling is important for the function of neurons as relay or storage devices of information in the brain. It has become increasingly clear that in the central nervous system, receptor activation and the coupled intracellular signaling, particularly that activated by growth factor receptors, is also required to sustain neuronal viability. Neurons are postmitotic cells, with a high degree of plasticity at synaptic junctions. The neuron therefore needs to maintain a high catabolic and anabolic capacity for remodeling and replenishing processes. These processes are stimulated by growth factors and also by external trophic influences such as neuronal activity, including membrane depolarization induced either by potassium

Laboratory for Experimental Brain Research, Wallenberg Neuroscience Center, Lund University Hospital, 22185 Lund, Sweden

Maturation Phenomenon in Cerebral Ischemia II
U. Ito et al. (Eds.)

or glutamate receptor activation [23]. When these stimuli are withdrawn, the neurons die, often through physiological cell death or programmed cell death and apoptosis [7, 13]. More recently it was shown that particular signaling pathways in neurons may activate neuronal death, in analogy with Fas receptor induced apoptosis in the immune system [12, 33].

For both intracellular communication and for cell survival and death the cellular signaling is crucial in transmitting information to the nucleus and to other cell organelles which could result in a transient or persistent change in cell function. The intracellular signals may be of different origin. Often a receptor is activated causing a subsequent opening of a coupled ion channel, but signals may also originate from mitochondria or endoplasmic reticulum. Elevation of calcium ion concentrations, be it through receptor-operated channels such as the NMDA receptor or released from intracellular stores such as mitochondria or endoplasmic reticulum, in turn activates subsequent reaction cascades (Fig. 1). Among these processes are protein phosphorylation-dephosphorylation reactions mediated by protein kinases which phosphorylate proteins or by protein phosphatases which dephosphorylate already phosphorylated proteins. Protein kinase C (PKC) and the calcium-calmodulin depen-

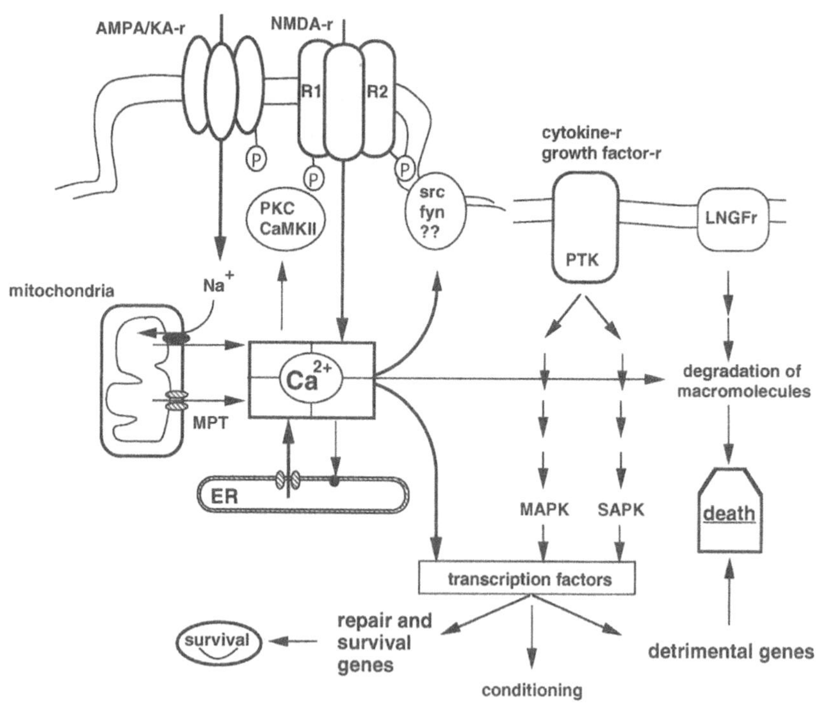

Fig. 1. Schematic drawing of intracellular signaling. *AMPA-r*, Amino-3-hydroxy-5-methyl-4-isoxazoleprionic acid receptor; *CaMKII*, calcium-calmodulin kinase II; *ER*, endoplasmic reticulum; *LNGFr*, low-affinity nerve growth factor receptor; *NMDA-r*, *N*-methyl-D-asparate receptor activation; *MAPK*, mitogen-activated protein kinase; *MPT*, mitochondrial phase transition; *PKC*, protein kinase C; *PTK*, protein tyrosine kinase; *SAPK*, stress-activated protein kinase

dent protein kinase II (CaMKII) are protein kinases involved in neurotransmission [22, 25] and memory [35]. These proteins are translocated from the soluble form (cytosolic) to the membrane-bound form upon cellular activation [4]. Alternatively, protein kinases are linked directly to the receptor protein such as the growth factor receptor protein tyrosine kinase, which in turn activates a cascade of protein kinases. Such cascades includes the mitogen-activated protein kinase II and the stress-activated protein kinase [37], which ultimately phosphorylate target transcription factors linking the plasma membrane and nucleus (Fig. 1). The latter pathway has been implicated in receptor-activated cell death. Recently it was demonstrated that the low-affinity nerve growth factor receptor, or p75, is structurally similar to the tumor necrosis factor-α receptor, suggesting that the receptor may transmit a direct death signal through the FADD/ICE system when not occupied by nerve growth factor [12, 33].

A short period of sublethal stress such as global or focal ischemia [6, 18, 19]; or spreading depression [16], induced several days prior to a lethal ischemic episode renders the brain resistant to subsequent period of lethal stress, a phenomenon called ischemic tolerance. Selective neuronal damage is induced in the brain, particularly in the hippocampal CA1 region, following transient cerebral ischemia [11], rat [32], and man [31]. The neuronal degeneration develops slowly during reperfusion following ischemia and is evident after 2–3 days of reperfusion [17, 32]. We have recently established a model of ischemic tolerance using the two-vessel occlusion model of ischemia. In these models a preconditioning ischemia of 3 min duration and with an interval of 2–3 days between the sublethal and lethal ischemic insults provides the maximal neuroprotective effect (Shamloo and Wieloch, unpublished).

Effect of Preconditioning on Cell Signaling

Ischemia activates and disturbs intracellular signaling [40]. In resistant brain regions the disturbance is normalized during recovery following ischemia but seems to persist in more vulnerable brain regions such as the CA1 region. In vulnerable regions, however, the distribution of protein kinases and the balance between phosphorylation and dephosphorylation is disturbed, and a persistent change in signaling is seen which ultimately may cause cell death. During the ischemic insult ATP decreases, and proteins are dephosphorylated.

During this period PKC and CaMKII are translocated from a soluble form to the plasma membrane, forming a strong association with the lipid bilayer. Figure 2 shows a Western blot of cortical homogenates separated into a cytosolic and a synaptosomal membrane fraction and stained with an antibody directed to the α-subunit of CaMKII, in accord with our previous results [10]. The levels of CaMKII in the cytosol decrease following ischemia and remain diminished for up to 24 h of reperfusion. On the other hand, the level increases in the synaptosomal fraction. Similar effects have been seen with regard to PKC [39]. The data suggest that a continuous stimulus is present during reperfusion which keeps PKC and CaMKII translocated to the plasma membrane. The identity of the stimulus is at present not known. It may be due to a persistent increase in calcium concentrations due to NMDA receptor activation or to a sustained release from intracellular stores such as the mitochondria or the endoplas-

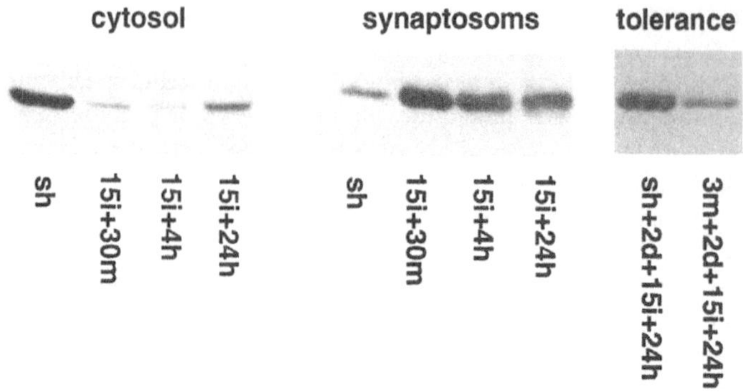

Fig. 2. The effect of transient ischemia on the levels of CaMKII in the cytosolic and membrane fraction of rats exposed to 15 min of transient ischemia and various period of recovery (*right*). The effect of preconditioning the rat brain by 3 min of ischemia followed by 2 days of recovery and a subsequent 15 min of ischemia and 2 days of recovery (*left*)

mic reticulum. The latter explanation posits that the opening state of the NMDA receptor is prolonged, or that mitochondria leaks calcium ions either through the mitochondrial phase transition pore [2, 27] or through exchange mechanisms such as the sodium-calcium antiporter. Tolerance markedly diminishes the membrane bound CaMKII levels (Fig. 2). In cortex from brains exposed to 3 min of ischemia and 2 days of recovery with as subsequent 15 min of ischemia and 24 h of recovery the levels of CaMKII in the membrane fraction were similar to those in the sham-operated animals. If translocation is regulated by the intracellular calcium concentrations, tolerance evidently enhances the normalization of calcium homeostasis following cerebral ischemia. An enhanced intracellular release of calcium ions could occur through the intracellular stores or through receptor coupled or voltage regulated channels. Tolerance prevents this calcium influx and secondarily inhibits the translocation of CaMKII and PKC.

Tyrosine phosphorylation is markedly enhanced following transient cerebral ischemia [9], particularly phosphorylation of the 180-kDa protein identified as the NR2 receptor protein in synaptic junction [26] (Fig. 3). We have earlier demonstrated that the phosphorylation of this protein is enhanced following ischemia (Hu and Wieloch, unpublished). This phosphorylation enhanced calcium conductance through the NMDA receptor operated channel [38]. Tolerance markedly affected tyrosine phosphorylation. The levels of phosphorylation were decreased by the preconditioning stimulus (Fig. 3). This was due partly to a decrease in the levels of the NR2A/B subunit (Fig. 3), where a clear decrease in the levels of the NR2A/B protein in the synaptosomal fraction is demonstrated. Preconditioning evidently activates factors that decrease the levels of the NR2 subunit as well as the tyrosine phosphorylation of the protein. The downregulation of the NR2 subunit and the normalization of the tyrosine phosphorylation could decrease the calcium flux through the NMDA receptor and therefore normalize intracellular calcium levels and diminish the subsequent translocation of CaMKII.

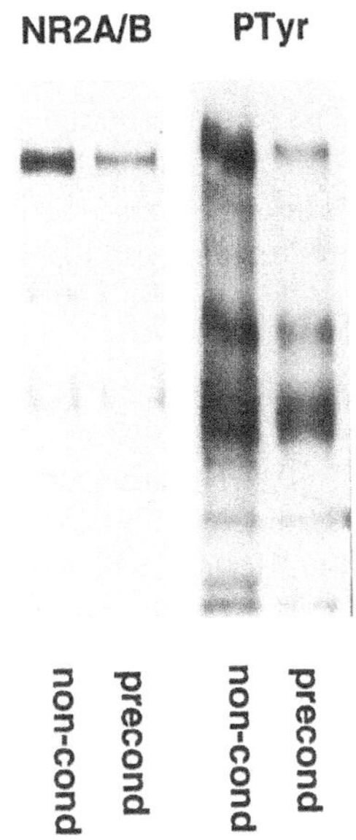

Fig. 3. The effect of preconditioning with 3 min of ischemia followed by 2 days of recovery, on the levels of the NMDA receptor subunit 2 and its tyrosine phosphorylation measured 24 h following 15 min of ischemia in the rat cortex

Development of Ischemic Tolerance

Since tolerance affects PKC, CaMKII, and the NMDA receptor protein levels, induction of tolerance may follow similar principles as seen in long-term potentiation or depression. We therefore envisage that tolerance is induced by a series of events: an *activation phase* occurring during the preconditioning insult, followed by an *induction phase* between the preconditioning and lethal ischemia where the tolerance inducing factor is established, and the *expression phase* during which tolerance is expressed (Fig. 4).

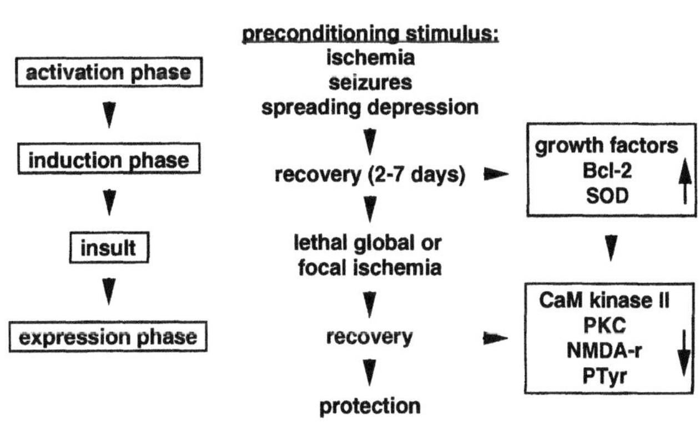

Fig. 4. Schematic drawing of the development of ischemic tolerance

Activation Phase

The tolerance-inducing event such as ischemia or spreading depression have some event in common that activates processes which induce tolerance. Receptor activation is important since blockade of the NMDA receptors as well as blockers of the adenosine receptor and the potassium channel blockers diminish tolerance [8, 14].

Induction Phase

Tolerance factors induced by the preconditioning stimulus are apparently established during the induction phase. It is likely that receptor stimulation activates transcriptions factors, which in turn stimulate the synthesis of tolerance factor(s).

The heat shock proteins, particularly HSP70, have been implicated in the induction of tolerance [3, 36], but contradictory results leaves this hypothesis unconfirmed [1, 20]. Enhanced expression of brain-derived neurotrophic factor [15] and the antiapoptotic protein bcl-2 [34] also occur during the induction phase. Since these proteins are neuroprotective, they may contribute to tolerance. Preconditioning facilitates recovery of protein synthesis following the second ischemic insult [28]. This suggests that factors enhancing protein synthesis recovery are important for tolerance-induced protection.

Expression Phase

Membranes are depolarized during the lethal ischemic insult, but we found no difference in the time to depolarization between conditioned and nonconditioned animals (Shamloo, unpublished). Also, the release of glutamate during ischemia is not affected by preconditioning [29].

As seen in Figs. 2 and 3, we observed a dramatic decrease in CaMKII in the membrane fraction and the level of NMDA receptor subunits and a decrease in tyrosine phosphorylation of the NR2 subunit at 24 h of recovery. One possible mechanism by which NMDA receptors are downregulated is by a cytokine-mediated mechanism [5, 24, 36]. The downregulation of CaMKII could also occur through a similar mechanism. Alternatively, the downregulation of NR2 leads to attenuation of intracellular calcium levels causing less CaMKII to be translocated and less NR2 to be tyrosine phosphorylated.

Conclusions

We propose that tolerance is established by an initial activation of gene programs that induce neuroprotective factors such as neurotrophins or cytokines in neurons and/or glia cells. The growth factors stimulate protein synthesis and the formation of neuroprotective Bcl-2. On the other hand, downregulated are proteins involved in neuronal activation, including protein kinases, receptor and membrane proteins such as the NMDA receptor complex, and presumably proteins in mitochondria and endoplasmic reticulum.

Acknowledgments. This work was supported by the Swedish Medical Research Council (grant 8644), the European Union (BIOMED II grant BMH4-CT96-0851), United States Public Health Services (grant NS7838), the Medical Faculty at University of Lund, and the Bergendahls Foundation.

References

1. Abe H, Nowak T (1995) An optimized model of induced ischemic tolerance in the Gerbil. Soc Neurosci Abs 21: 1728
2. Ankarcrona M, Dypbukt JM, Bonfoco E, Zhivotovsky B, Orrenius S, Lipton SA, Nicotera P (1995) Glutamate-induced neuronal death: a succession of apoptosis and necrosis depending on mitochondrial function. Neuron 15: 961–973
3. Aoki M, Abe K, Kawagoc J, Nakamura S, Kogure K (1993) Acceleration of HSP70 and HSC70 heat shock gene expression following transient ischemia in the preconditioned gerbil hippocampus. J Cereb Blood Flow Metab 13: 781–788
4. Bronstein JM, Farber DB, Wasterlain CG (1993) Regulation of type II calmodulin kinase: function implications. Brain Res Rev 18: 135–147
5. Cheng B, Christakos S, Mattson MP (1994) Tumor necrosis factors protect neurons against metabolic-excitotoxic insults and promote maintenance of calcium homeostasis. Neuron 12: 139–153
6. Glazier SS, O'Rourke DM, Graham DI, Welsh FA (1994) Induction of ischemic tolerance following brief focal ischemia in rat brain. J Cereb Blood Flow Metab 14: 545–553
7. Hale AJ, Smith CA, Sutherland LC, Stoneman VEA, Longthorne VL, Culhane AC, Williams GT (1996) Apoptosis: molecular regulation of cell death. Eur J Biochem 236: 1–26
8. Heurteaux C, Lauritzen I, Widmann C, Lazdunski M (1995) Essential role of adenosine, adenosine A1 receptors, and ATP-sensitive K^+ channels in cerebral ischemic preconditioning. Proc Natl Acad Sci USA 92: 4666–4670
9. Hu BR, Wieloch T (1994) Tyrosine phosphorylation and activation of MAP kinase in the rat brain following transient cerebral ischemia. J Neurochem 62: 1357–1367
10. Hu BR, Wieloch T (1995) Persistent translocation of Ca^{2+}/calmodulin-dependent protein kinase II to synaptic junctions in the vulnerable hippocampal CA1 region following transient ischemia. J Neurochem 64: 277–284
11. Ito U, Spatz M, Walker JT Jr, Klatzo I (1975) Experimental cerebral ischemia in Mongolian gerbils. I. Light microscopic observations. Acta Neuropathol Berl 32: 209–223
12. Itoh N, Nagata S (1993) A novel protein domain required for apoptosis. J Biol Chem 268: 10932–10937
13. Johnson EM, Deckwerth TL (1993) Molecular mechanisms of developmental neuronal death. Ann Rev Neurosci 16: 31–46
14. Kato H, Liu Y, Araki T, Kogurc K (1992) MK-801, but not anisomycin, inhibits the induction of tolerance to ischemia in the gerbil hippocampus. Neurosci Lett 139: 118–121
15. Kawahara N, Reutzler C, Wiegand S, Croll S, Klatzo I (1994) Induction of protective mechanisms in cerebral ischemia. Soc Neurosci Abs 20: 1480
16. Kawahara N, Ruetzler CA, Klatzo I (1995) Protective effect of spreading depression against neuronal damage following cardiac arrest cerebral ischaemia. Neurol Res 17: 9–16
17. Kirino T (1982) Delayed neuronal death in the gerbil hippocampus following ischemia. Brain Res 239: 57–69
18. Kirino T, Tsujita Y, Tamura A (1991) Induced tolerance to ischemia in gerbil hippocampal neurons. J Cereb Blood Flow Metab 11: 299–307
19. Kitagawa K, Matsumoto M, Tagaya M, Hata R, Ueda H, Niinobe M, Handa N, Fukunaga R, Kimura K, Mikoshiba K (1990) "Ischemic tolerance" phenomenon found in the brain. Brain Res 528: 21–24

20. Kobayashi S, Harris VA, Welsh FA (1995) Spreading depression induces tolerance of cortical neurons to ischemia in rat brain. J Cereb Blood Flow Metab 15: 721–727
21. Kristain T, Siesjö BK (1996) Calcium-related damage in ischemia. Life Sciences 59: 357–367
22. Llinás R, Gruner J-A, Sugimori M, McGuinness T, Greengard P (1991) Regulation by synapsin I and calcium/calmodulin dependent protein kinase II of transmitter release at the squid gaint synapse. Proc Natl Acad Sci USA 82: 3035–3039
23. Mattson MP, Murrain M, Gurthrie PB, Karter SB (1989) Fibroblast growth factor and glutamate: opposing roles in the generation and degeneration of hippocampal neuroarchitecture. J Neuroscience 9: 3728–3741
24. Mattson MP, Kumar KN, Wang H, Cheng B, Michaelis EK (1992) Basic FGF regulates the expression of a functional 71 kDa NMDA receptor protein that mediates calcium influx and neurotoxicity in hippocampal neurons. J Neurosci 13: 4575–4588
25. McGlade-McCulloh E, Yamamoto H, Tan SE, Brickey DA, Soderling TR (1993) Phosphorylation and regulation of glutamate receptors by calcium/calmodulin-dependent protein kinase II. Nature 362: 640–642
26. Moon IS, Apperson ML, Kennedy MB (1994) The major tyrosine-phosphorylated protein in the postsynaptic density fraction is *N*-methyl-D-aspartate receptor subunit 2B. Proc Natl Acad Sci USA 91: 3954–3958
27. Mulkey RM, Endo S, Shenolikar S, Malenka RC (1994) Involvement of a calcineurin/inhibitor-1 phosphatase cascade in hippocampal long-term depression. Nature 369: 486–488
28. Nakagomi T, Kirino T, Kanemitsu H, Tsujita Y, Tamura A (1993) Early recovery of protein synthesis following ischemia in hippocampal neurons with induced tolerance in the gerbil. Acta Neuropathol (Berl) 86: 10–15
29. Nakata N, Kato H, Liu Y, Kogure K (1992) Effects of pretreatment with sublethal ischemia on the ectracellular glutamate concentrations during secondary ischemia in the gerbil hippocampus evaluated with intarcerebral microdialysis. Neurosci Lett 138: 86–88
30. Nishi S, Taki W, Uemura Y, Higashi T, Kikuchi H, Kudoh H, Satoh M, Nagata K (1993) Ischemic tolerance due to the induction of HSP70 in a rat ischemic recirculation model. Brain Res 615: 281–288
31. Petito CK, Feldmann E, Pulsinelli WA, Plum F (1987) Delayed hippocampal damage in humans following cardiorespiratory arrest. Neurology 37: 1281–1286
32. Pulsinelli WA, Brierley JB, Plum F (1982) Temporal profile of neuronal damage in a model of transient forebrain ischemia. Ann Neurol 11: 491–498
33. Rabizadeh S, Oh J, Zhong L, Yang J, Bitler CM, Butcher LL, Bredesen DE (1993) Induction of apoptosis by the low-affinity NGF receptor. Science 261: 345–348
34. Shimazaki K, Ishida A, Kawai N (1994) Increase in bcl-2 oncoprotein and the tolerance to ischemia-induced neuronal death in the gerbil hippocampus. Neurosci Res 20: 95–99
35. Silva AJ, Stevens CF, Tonegawa S, Wang Y (1992) Deficient hippocampal long-term potentiation in alpha-calcium-calmodulin kinase II mutant mice. Science 257: 201–206
36. Strijbos PJ, Rothwell NJ (1995) Interleukin-1 beta attenuates excitatory amino acid-induced neurodegeneration in vitro: involvement of nerve growth factor. J Neurosci 15: 3468–3474
37. Verheij M, Bose R, Lin XH, Yao B, Jarvis WD, Grant S, Birrer MJ, Szabo E, Zon LI, Kyriakis JM, Haimovitz, Friedman A, Fuks Z, Kolesnick RN (1996) Requirement for ceramide-initiated SAPK/JNK signalling in stress-induced apoptosis. Nature 380: 75–79
38. Wang YT, Salter MW (1994) Regulation of NMDA receptors by tyrosine kinases and phosphatases. Nature 369: 233–235
39. Wieloch T, Cardell M, Bingren H, Zivin J, Saitoh T (1991) Changes in the activity of protein kinase C and the differential subcellular redistribution of its isozymes in the rat striatum during and following transient forebrain ischemia. J Neurochem 56: 1227–1235
40. Wieloch T, Hu B, Boris-Möller F, Cardell M, Kamme F, Kurihara J, Sakata K (1996) Intracellular signal transduction in the postischemic brain. In: Sjesjö BK, Wieloch T (eds) Advances in neurology; cellular and molecular mechanisms of ischemic brain damage. Lippincott-Raven, Philadelphia, pp 371–388 (Advances in neurology, vol 71)

Electrophysiological Evaluation of Ischemic Tolerance Phenomenon

K. KAWAI[1,2], T. NAKAGOMI[1], T. KIRINO[2], A. TAMURA[1], and N. KAWAI[3]

Summary

We investigated electrophysiological aspects of ischemic tolerance phenomenon in gerbil hippocampus using extracellular recordings from hippocampal slice preparations. Firstly, the capacity for long-term potentiation triggered by tetanic stimulation, which is a putative synaptic background of memory, was examined in hippocampal slices from gerbils subjected to double ischemia. Capacity for long-term potentiation was transiently inhibited 2 days following double ischemia but then recovered 7 days following double ischemia, suggesting that preconditioning ischemia functionally protected CA1 neurons. Secondly, the latency of anoxic depolarization and postanoxic potentiation of *N*-methyl-D-aspartate (NMDA) receptor-mediated transmission was examined by subjecting hippocampal slices to in vitro anoxia. Latency of anoxic depolarization was not significantly different between slices from gerbils 2 days after sham operations and those from gerbils 2 days after preconditioning ischemia. Postanoxic potentiation of NMDA receptor-mediated transmission was not observed in slices from gerbils 2 days following preconditioning ischemia, while it was observed in slices from gerbils 2 days following sham operations or 9 days following preconditioning ischemia, suggesting that inhibition of anoxic potentiation of NMDA receptor-mediated transmission may be one of the mechanisms by which induced tolerance protects CA1 neurons.

Introduction

Pyramidal neurons in the hippocampal CA1 sector are vulnerable to cerebral ischemia and show delayed neuronal death 2–4 days after transient ischemia [11, 18]. It was recently reported that brief ischemia, which is not in itself lethal to neurons, induces tolerance to subsequent ischemic episodes in hippocampal CA1 neurons in the gerbil [12, 14] and rat [15]. Mechanisms underlying this ischemic tolerance phenomenon remain unknown, although indirect evidence suggests involvement of induction of

[1] Department of Neurosurgery, Teikyo University School of Medicine, 2-11-1 Kaga, Itabashi, Tokyo 173, Japan
[2] Department of Neurosurgery, Faculty of Medicine, University of Tokyo, 7-3-1 Hongo, Bunkyo-ku, Tokyo 113, Japan
[3] Department of Physiology, Jichi Medical School, 3311-1 Yakushiji, Minamikawachi-machi, Tochigi 329-04, Japan

Maturation Phenomenon in Cerebral Ischemia II
U. Ito et al. (Eds.)

protective proteins and enzymes, such as heat shock proteins and superoxide dismutases [1, 8–10, 12, 16, 17]. It is also undetermined whether CA1 neurons are functionally protected by the acquired tolerance, since in previous studies ischemic tolerance has been evaluated only morphologically. The purpose of this study is to investigate electrophysiologically the functional aspects of ischemic tolerance phenomenon using hippocampal slices from gerbils subjected to single non-lethal ischemia or double ischemia. In this study, functional protection has been evaluated as the capacity for long-term potentiation (LTP) of synaptic transmission induced by brief high-frequency stimulation, since LTP is considered to be a putative synaptic background of memory. To investigate changes in electrophysiological properties which may underlie the protective mechanisms of ischemic tolerance, we have examined the latency of anoxic depolarization and then LTP of *N*-methyl-D-aspartate (NMDA) receptor-mediated synaptic transmission induced by in vitro anoxia (anoxic LTP) in slices from preconditioned gerbils.

Materials and Methods

In Vivo Preparation

Male Mongolian gerbils weighing 60–80 g were subjected to preconditioning ischemia (2 min) in an awake condition using Tone's method [20]. A nylon thread, which had been looped around the carotid arteries on the previous day, was pulled and the carotid arteries were occluded bilaterally. The thread was only removed in the sham-operated group. In the double ischemia group, gerbils were anesthetized with halothane 2 or 3 days after the preconditioning ischemia and the carotid arteries were exposed and occluded for 5 min with Sugita aneurysm clips. During the operation, the cranial temperature was regulated at 37.5°C ± 0.3°C using a heating pad and lamp.

Slice Preparation

Transverse sections (300–400 μm) were prepared by a previously described method [13, 21]. The slices were maintained at 34°C–35°C, and were perfused (0.5–1 ml/min) with an artificial cerebrospinal fluid (ACSF) containing 124 m*M* NaCl, 26 m*M* $NaHCO_3$, 5 m*M* KCl, 1.24 m*M* KH_2PO_4, 2.4 m*M* $CaCl_2$, 1.3 m*M* $MgSO_4$, and 10 m*M* glucose, equilibrated with 95% O_2 and 5% CO_2. Slices were kept in the chamber for at least 2 h before experiments. Field excitatory postsynaptic potentials (fEPSPs) were recorded extracellularly from the stratum radiatum of the CA1 region using glass electrodes filled with saline (1–2 MΩ). In some experiments, direct current (DC) potentials were recorded simultaneously using the same electrodes. Schaffer collateral–commissural afferents were stimulated (0.05–0.08 Hz) with bipolar platinum–iridium electrodes placed in the stratum radiatum. For experiments of anoxic depolarization and of anoxic long-term potentiation, slices were subjected to in vitro anoxia. The perfusate was transiently changed to the anoxic–aglycemic medium containing 10 m*M* sucrose instead of 10 m*M* glucose, equilibrated with 95% N_2 and 5% CO_2.

Tetanic Long-Term Potentiation

Hippocampal slice preparations were made from normal control gerbils, gerbils 2–3 days after the preconditioning ischemia, and gerbils 2 days or 7 days after double ischemia. Tetanic stimulation consisted of two trains of 100 pulses at 100 Hz with an interval of 20 s. To measure the maximal rate of rise of fEPSPs, records were digitized at 10 kHz and analyzed by an on-line computer. The computed slope values were displayed as a constantly updated time series, permitting the time course of synaptic potentiation to be monitored accurately during experiments.

Latency of Anoxic Depolarization

Hippocampal slice preparations were made from gerbils 2 days after sham operations and gerbils 2 days after the preconditioning ischemia. Two slices from each group were placed in a single chamber and DC potentials were continuously recorded from both slices. After changing the perfusate to the anoxic–aglycemic medium, DC potentials suddenly decreased by 5–10 mV in most slices (anoxic depolarization). The latencies from the start of in vitro anoxia to anoxic depolarization in both slices were recorded. The perfusate was changed back to the normal ACSF a few minutes after anoxic depolarization was observed in both slices or otherwise 10 min after the start of in vitro anoxia.

Anoxic Long-Term Potentiation

Hippocampal slice preparations were made from normal control gerbils, gerbils 2–3 days after sham operations, and gerbils 2–3 days or 9 days after the preconditioning ischemia. Following perfusion with ACSF for 2 h, the CA1 region was isolated from the CA3 by a knife cut to prevent spread of bursting activity from the latter regions. Slices were then perfused for at least 30 min with a modified ACSF containing CNQX (15 μM) and a lower $MgSO_4$ concentration (0.1 mM) to completely inhibit DL-α-amino-3-hydroxy-5-methyl-4-isoxazole propionic acid (AMPA) receptor responses and thus allow us to observe only NMDA receptor-mediated transmission. After obtaining a stable series of fEPSPs, slices were subjected to in vitro anoxia for 2–3 min. Monitoring of fEPSPs was continued for at least 30 min thereafter.

Results

Tetanic Long-Term Potentiation

In hippocampal slices from normal control gerbils, tetanic stimulation of Schaffer collateral–commissural fibers induced a potentiation of fEPSPs. Long-lasting synaptic potentiation was manifested as an increase in maximal rate of the initial rise of fEPSPs, which was more than 150% and lasted for at least 30 min in most slices (Fig. 1A). Similar potentiation of fEPSPs induced by tetanic stimulation was observed in

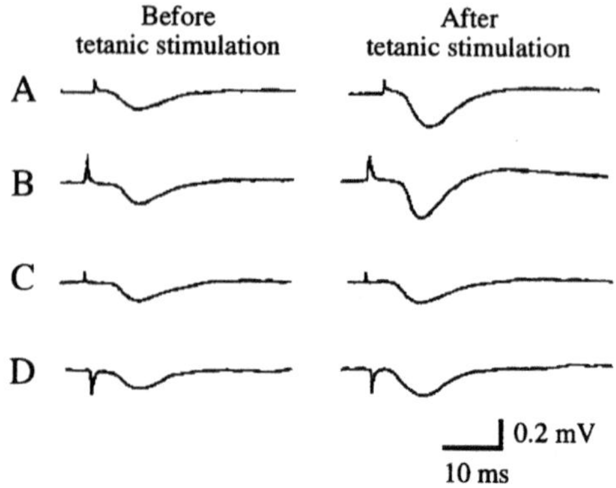

Fig. 1A–D. Field excitatory postsynaptic potentials recorded from hippocampal slices before (*left*) and 30 min after (*right*) tetanic stimulation. **A** Normal control gerbil. **B** Gerbil 2 days following preconditioning ischemia. **C** Gerbil 2 days following double ischemia. **D** Gerbil 7 days following double ischemia

most slices from gerbils 2 days after the preconditioning ischemia (Fig. 1B). In contrast, in most (>80%) slices from gerbils 2 days after the double ischemia, tetanic stimulation failed to induce robust potentiation of fEPSP. The increase in maximum rate of the initial rise of fEPSPs in these slices was, if any, less than 150% and did not last as long as that observed in slices from normal control gerbils (Fig. 1C). However, approximately 60% of slices from gerbils 7 days after the double ischemia again manifested robust potentiation of fEPSPs following tetanic stimulation (Fig. 1D).

Latency of Anoxic Depolarization

In most hippocampal slices, anoxic depolarization was observed within 5 min after initiation of in vitro anoxia. There was no tendency for slices from either the sham-operated or preconditioned gerbils to exhibit depolarization earlier. The mean latencies of anoxic depolarization were 2.88 ± 1.90 min and 3.00 ± 1.85 min in the sham-operated group and in the preconditioned group, respectively, which were not significantly different between groups.

Anoxic Long-Term Potentiation

Changing the perfusate from normal ACSF to modified ACSF containing CNQX and a low concentration of magnesium resulted in NMDA-mediated fEPSPs which were characterized by a long duration with a slower decay by inhibition of the AMPA component [4]. In slices from normal control gerbils, 2–3 min of in vitro anoxia induced a sudden decrease and then disappearance of fEPSPs. Changing the medium back to normal ACSF resulted in a gradual recovery and then potentiation of fEPSPs (anoxic LTP). Similar anoxic LTP was observed in slices from gerbils 2–3 days after sham operation and gerbils 9 days after the preconditioning ischemia (Fig. 2A, B, D). In contrast, no slices from gerbils 2–3 days after the preconditioning ischemia manifested anoxic LTP. Although fEPSPs gradually recovered after in vitro anoxia in these slices, they never reached the level attained before in vitro anoxia (Fig. 2C).

Discussion

Although the ability to exhibit LTP following tetanic stimulation in the hippocampus was inhibited 2 days after double ischemia, it was recovered 7 days after double ischemia. In contrast, our previous study revealed that a majority of CA1 neurons lost the capacity for LTP continuously following single 5 min ischemia [13]. The fact that the recovery in the capacity for LTP 7 days after double ischemia was not perfect but was observed in approximately 60% of slices is consistent with the result of our previous morphological study using the same gerbil model of forebrain ischemia, which demonstrated that approximately 50% of the CA1 neurons were preserved, when evaluated 7 days after double ischemia in which the interval of two ischemic insults was 2 days [12]. This study demonstrated that the protection afforded by ischemic preconditioning is not only morphological but may also be functional since growing evidence has shown that LTP underlies at least certain forms of memory [2].

Neurons in the hippocampal CA1 area slightly hyperpolarize initially during a short period of hypoxia, which are most likely mediated by activation of K^+ channels. With longer periods of hypoxia, the small hyperpolarization is followed by a major and rapid depolarization, which is called anoxic depolarization [7]. In the present study we detected no difference in latency from the start of in vitro anoxia to anoxic depolarization between hippocampal slices from gerbils 2 days after preconditioning

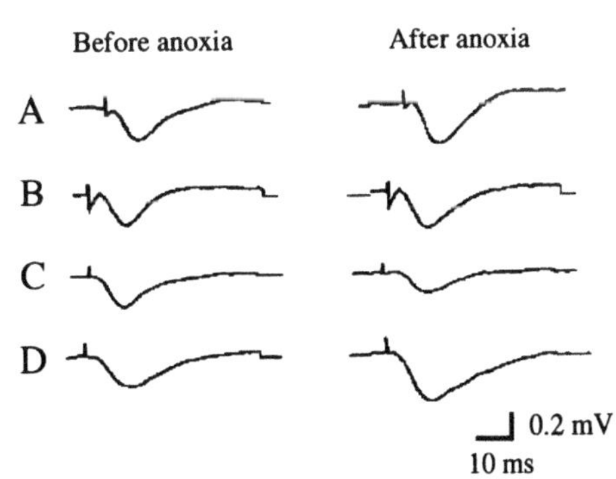

Fig. 2A–D. Field excitatory postsynaptic potentials recorded from hippocampal slices perfused with medium containing CNQX and a low concentration of magnesium. Records before and 30 min after 2–3 min of in vitro anoxia are presented on the *left* and on the *right*, respectively. **A** Normal control gerbil. **B** Gerbil 2 days following sham operation. **C** Gerbil 2 days following preconditioning ischemia. **D** Gerbil 9 days following preconditioning ischemia

ischemia and from gerbils 2 days after sham operations. This suggests that modification of cellular events, which occur earlier than anoxic depolarization during ischemia, may not be involved in the mechanisms of protection by preconditioning.

What then is the underlying mechanism of ischemic tolerance phenomenon? Induction of protective proteins, such as heat shock protein (HSP)70, ubiquitin and Bcl-2, and induction of protective enzymes such as superoxide dismutase have been suggested [1, 8–10, 12, 16, 17, 19]. Our result raises another possible mechanism, namely redox modulation of NMDA receptors [6]. Crepel et al. first reported that an anoxic–aglycemic episode induced a selective long-term potentiation of NMDA receptor-mediated fEPSPs and currents, suggesting that this potentiation could play an important role in ischemic-delayed neuronal death [3, 4]. Subsequently, the same group reported that the redox site of the NMDA receptor is involved in the induction and maintenance of the anoxic LTP of the NMDA receptor-mediated response [5]. In the present study we have demonstrated that slices from gerbils 2 days after preconditioning ischemia did not exhibit anoxic LTP, while anoxic LTP was observed in slices from gerbils 2 days after sham operations and gerbils 9 days after preconditioning ischemia. Although the cause–effect relationship of this inhibited anoxic LTP in preconditioned gerbils and the protective effect of preconditioning ischemia is unclear, decreased NMDA receptor-mediated transmission by modulation of the redox site of receptors following brief non-lethal ischemia may play an important role in the mechanism of ischemic tolerance phenomenon.

References

1. Aoki M, Abe K, Kawagoe J, Nakamura S, Kogure K (1993) The preconditioned hippocampus accelerates HSP70 heat shock gene expression following transient ischemia in the gerbil. Neurosci Lett 155: 7–10
2. Bliss TVP, Collingridge GL (1993) A synaptic model of memory: long-term potentiation in the hippocampus. Nature 361: 31–39
3. Crepel V, Hammond C, Krnjevic K, Chinestra P, Ben-Ari Y (1993a) Anoxia-induced LTP of isolated NMDA receptor-mediated synaptic responses. J Neurophysiol 69: 1774–1778
4. Crepel Y, Hammond C, Chinestra P, Diabira D, Ben-Ari Y (1993b) A selective LTP of NMDA receptor-mediated currents induced by anoxia in CA1 hippocampal neurons. J Neurophysiol 70: 2045–2055
5. Gozlan H, Ben-Ari Y (1995) NMDA receptor redox sites: are they targets for selective neuronal protection? TIPS 16: 368–374
6. Gozlan H, Diabira D, Chinestra P, Ben-Ari Y (1994) Anoxic LTP is mediated by the redox modulatory site of the NMDA receptor. J Neurophysiol 72: 3017–3022
7. Haddad GG, Jiang C (1993) O_2 deprivation in the central nervous system: on mechanisms of neuronal response, differential sensitivity and injury. Prog Neurobiol 40: 277–318
8. Kato H, Chen T, Liu X-H, Nakata N, Kogure K (1993) Immunohistochemical localization of ubiquitin in gerbil hippocampus with induced tolerance to ischemia. Brain Res 619: 339–343
9. Kato H, Liu Y, Kogure K, Kata K (1994) Induction of 27-kDa heat shock protein following cerebral ischemia in a rat model of ischemic tolerance. Brain Res 634: 235–244
10. Kato H, Kogure K, Araki T, Liu X-H, Kato K, Itoyama Y (1995) Immunohistochemical localization of superoxide dismutase in the hippocampus following ischemia in gerbil model of ischemic tolerance. J Cereb Blood Flow Metab 15: 60–70
11. Kirino T (1982) Delayed neuronal death in the gerbil hippocampus following ischemia. Brain Res 239: 57–69

12. Kirino T, Tsujita Y, Tamura A (1991) Induced tolerance to ischemia in gerbil hippocampal neurons. J Cereb Blood Flow Metab 11: 299–307
13. Kirino T, Robinson HPC, Miwa A, Tamura A, Kawai N (1992) Disturbance of membrane function preceding ischemic delayed neuronal death in the gerbil hippocampus. J Cereb Blood Flow Metab 12: 408–417
14. Kitagawa K, Matsumoto M, Tagaya M, Hata R, Ueda H, Niinobe M, Handa N, Fukunaga R, Kimura K, Mikoshiba K, Kamada T (1990) "Ischemic tolerance" phenomenon found in the brain. Brain Res 528: 21–24
15. Liu Y, Kato H, Nakata N, Kogure K (1992) Protection of rat hippocampus against ischemic neuronal damage by pretreatment with sublethal ischemia. Brain Res 586: 121–124
16. Liu Y, Kato H, Nakata N, Kogure K (1993) Temporal profile of heat shock protein 70 synthesis in ischemic tolerance induced by preconditioning ischemia in rat hippocampus. Neuroscience 56: 921–927
17. Ohtsuki T, Matsumoto M, Kuwabara K, Kitagawa K, Suzuki K, Taniguchi N, Kamada T (1992) Influence of oxidative stress on induced tolerance to ischemia in gerbil hippocampal neurons. Brain Res 599: 246–252
18. Pulsinelli WA, Brierley JB, Plum F (1982) Temporal profile of neuronal damage in a model of transient forebrain ischemia. Ann Neurol 11: 491–498
19. Shimazaki K, Watanabe K, Hosoya H, Oguro K, Masuzawa T, Eguchi Y, Tsujimoto Y, Kawai N (1995) Bcl-2 expression and tolerance for postischemic neuronal death in the gerbil hippocampus. Soc Neurosci Abstr 21: 1727
20. Tone O, Miller JC, Bell JM, Rapoport SI (1987) Regional cerebral palmitate incorporation following transient bilateral carotid occlusion in awake gerbils. Stroke 18: 1120–1127
21. Tsubokawa H, Oguro K, Robinson HPC, Masuzawa T, Kirino T, Kawai N (1992) Abnormal Ca^{2+} homeostasis before cell death revealed by whole cell recording of ischemic CA1 hippocampal neurons. Neuroscience 49: 807–817

Transient Ischemia Depletes Free Ubiquitin in the Gerbil Hippocampal CA1 Neurons

T. Ide, T. Morimoto, and T. Kirino

Summary

We investigated ubiquitin immunoreactivity in the postischemic gerbil hippocampus using two kinds of antibodies. Immunostaining for ubiquitin in the hippocampus was strongly dependent on the antibodies used. Using rabbit polyclonal antibody U-5379, immunoreactivity disappeared from the hippocampus in the early reperfusion period, and reappeared in the dentate granule cells and CA3 pyramidal cells, but never in the CA1 pyramidal cells. In contrast, rat monoclonal antibody DF2 showed sustained immunoreactivity in the CA1 during the 48-h reperfusion period. On the immunoblots of gerbil brain homogenates, U-5379 and DF2 exhibited similar specificities. Immunoprecipitation and immunoabsorption of these antibodies disclosed that, under nondenaturing conditions, U-5379 bound exclusively free ubiquitin, while DF2 had little affinity for free ubiquitin, but appeared to have more affinity for conjugated ubiquitin. Thus, transient ischemia depletes free ubiquitin but not conjugated ubiquitin in the CA1. This depletion may be caused by impaired conversion from conjugated to free ubiquitin and/or failure of de novo ubiquitin synthesis.

Introduction

Ubiquitin is a small protein, which is an essential factor for the ATP-dependent nonlysosomal protein degradation. Once ubiquitinated, target proteins are rapidly degraded by the 26S proteasome with broad specificity [5].

Magnusson and Wieloch [3] found a sustained loss of ubiquitin immunoreactivity in the CA1 neurons following brief forebrain ischemia in the rat. Since ubiquitin is one of the heat shock proteins [1], these findings were interpreted as indicating a defective stress response of CA1 neurons. Thus, it has been hypothesized that impairment of the stress response and a defective ubiquitin pathway in the CA1 bring about an accumulation of abnormal proteins and profound effects on the turnover of short-lived, regulatory proteins, which could play an essential role in delayed neuronal death following ischemia.

Here, we report the changes in ubiquitin immunoreactivity in the postischemic gerbil hippocampus as shown with two kinds of ubiquitin antibodies. Our results

Department of Neurosurgery, Faculty of Medicine, University of Tokyo 7-3-1 Hongo, Bunkyo-ku, Tokyo 113, Japan

Maturation Phenomenon in Cerebral Ischemia II
U. Ito et al. (Eds.)

indicate that free ubiquitin, but not conjugated ubiquitin, is depleted in CA1 neurons following ischemia.

Materials and Methods

Induction of Ischemia. Male Mongolian gerbils (*Meriones unguiculatus*, 60–80 g) were subjected to 5 min of forebrain ischemia. During the operation, the body (rectal) and temporal muscle temperatures were monitored and maintained at 37.5°C ± 0.4°C. At reperfusion intervals of 6 h, 24 h, and 48 h, the animals were fixed by transcardiac perfusion with 4% paraformaldehyde in 0.1 *M* phosphate buffer (pH 7.4), and vibratome sections containing the dorsal hippocampus were prepared and processed as described below.

Antibodies. Primary antibodies used in these studies were rabbit antiubiquitin antiserum U-5379 (Sigma, St. Louis, MO) and rat monoclonal antibody DF2 (IgM) [4]. DF2 was raised against paired helical filaments (PHF).

Immunocytochemistry. Immunocytochemical staining was performed using the Vectastain ABC kit (Vector Laboratories, Burlingame, CA). After destroying endogenous peroxidase and blocking, the sections were incubated with U-5379 at a dilution of 1:100 overnight at 4°C, followed by incubation with biotinylated goat anti-rabbit IgG and avidin-biotinylated horseradish peroxidase complex solution. Peroxidase was detected with 0.5 mg/ml 3,3′-diaminobenzidine tetrahydrochloride and 0.01% H_2O_2 in Tris-buffered saline (TBS).

Immunocytochemical staining with DF2 at a dilution of 1:1 (culture medium) was performed similarly with appropriate blocking and secondary antibody using rat ABC kit.

Immunoblotting. Male Mongolian gerbils were decapitated and the brain was removed and homogenized in three volumes of TBS containing 1 m*M* ethyleneglycoltetraacetic acid (EGTA) and proteinase inhibitors. Extracts with TBS, 1% Triton X-100, and 4 *M* urea, and urea-insoluble fraction were prepaired. A small aliquot from each extraction was diluted in Laemmli's sample buffer, boiled, and electrophoresed on 17.5% polyacrylamide gels. After electrotransfer to polyvinylidene difluoride (PVDF) membranes (Immobilon, Millipore, Bedford, MA), the membranes were incubated with two ubiquitin antibodies, followed by biotinylated secondary antibodies and avidin-biotinylated horseradish peroxidase complex. Bound antibodies were visualized after exposure to 0.5 mg/ml 4-chloro-l-naphthol and 0.015% H_2O_2 in methanol and TBS (1:5).

Immunoprecipitation. A male Mongolian gerbil brain was homogenized in nine volumes of TBS containing 1 m*M* EGTA and proteinase inhibitors. The homogenate was centrifuged. The supernatant was mixed with immobilized protein G (Pierce, Rockford, IL) and centrifuged to remove nonspecific binding. The supernatant was incubated with U-5379 diluted at 1:25 overnight at 4°C. The immune complexes were precipitated with 10% immobilized protein G. The pellet was washed five times with

TBS (pH 8.0) containing 0.1% sodium dodecyl sulphate (SDS), 0.5% deoxycholate, and 1% Nonidet P40, and suspended in Laemmli's sample buffer. The mixture was boiled, electrophoresed, and electrotransferred, and the blot was probed with DF2. The procedure for immunoblotting was the same as above. Immunoprecipitation with DF2 was performed in a similar manner with the same concentrations as those for immunocytochemistry. Electroblots were probed with DF2.

Immunoabsorption. Free ubiquitin (U-6253; Sigma, St. Louis, MO) was electrophoresed (~0.5 μg per lane), and the blot was probed with U-5379 and DF2 which had been incubated with various concentrations of free ubiquitin at 0, 20, 50, 100, 200, and 400 μg/ml, overnight at 4°C. The procedure for immunoblotting was the same as above. The concentration of each antibody was also the same as that for immunocytochemistry.

Results and Discussion

Immunocytochemistry. The ubiquitin immunoreactivity with U-5379 disappeared from the CA1 following ischemia (Fig. 1). At 6 h of reperfusion following 5-min ischemia, the immunoreactivity was undetectable in almost all the neurons except for interneurons in the hippocampus. At 24 h of reperfusion, the immunoreactivity in the dentate granule cells and CA3 pyramidal cells reappeared and returned to the control level at 48 h; however, it never reappeared in the CA1.

In contrast to U-5379, DF2 revealed sustained ubiquitin immunoreactivity in the CA1 (Fig. 1). Ubiquitin staining in all hippocampal neurons gradually increased at 24 h, and returned to the control level at 48 h.

Immunoblotting. U-5379 labeled predominantly free ubiquitin at 6 kDa in TBS-, Triton X-100-, and ureaextracted fractions. DF2 labeled strongly free ubiquitin and four other bands at 14, 17, 23 and 80 kDa. In addition to these definite bands, many high-molecular-weight bands were discernible. Although there are some differences in the specificities on the immunoblots among the two kinds of antibodies, these cannot account for the distinct differences in the immunostaining of the CA1 pyramidal cells following ischemia.

Immunoprecipitation. The DF2 immunoblot of the U-5379 immunoprecipitate showed free ubiquitin and many faintly stained high-molecular-weight bands (Fig. 2). The immunoprecipitates with DF2 contained a much smaller amount of free ubiquitin, whereas high-molecular-weight ubiquitin conjugates were somewhat abundant (Fig. 2).

Immunoabsorption. The blot staining with U-5379 disappeared after preabsorption with free ubiquitin at concentrations of more than 50 μg/ml, while that of DF2 remained the same after preabsorption. Thus, it is likely that U-5379 had higher affinity toward free ubiquitin than DF2 under nondenaturing conditions.

Immunostaining with U-5379 was almost identical with that reported by Magnusson and Wieloch [3]. In contrast, DF2 clearly showed sustained ubiquitin immunoreactivity during the 48-h reperfusion period. Since U-5379 preferentially recognizes free ubiquitin under nondenaturing conditions, immunocytochemical stainings with U-5379 indicate that free ubiquitin level is markedly reduced and never recovers during 48 h of postischemia in the postischemic CA1.

The second conclusion is that the level of conjugated ubiquitin is not greatly affected in the postischemic CA1. This is based on the immunostaining with DF2 which mainly recognizes conjugated ubiquitin under nondenaturing conditions. Con-

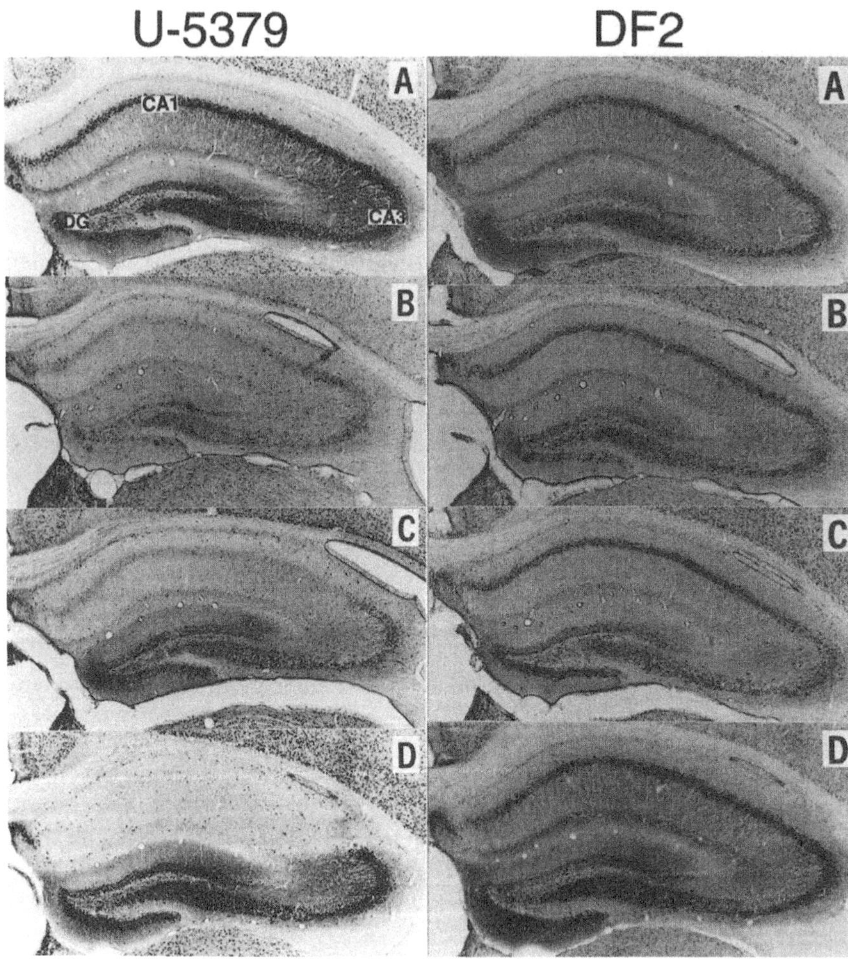

Fig. 1. Immunocytochemical staining for ubiquitin with rabbit polyclonal antibody U-5379 and rat monoclonal antibody DF2 in the hippocampus of control gerbils (**A**) and gerbils subjected to 5-min ischemia followed by 6 h (**B**), 24 h (**C**), and 48 h (**D**) of reperfusion. With U-5379, after disappearance at 6 h of reperfusion, the ubiquitin immunoreactivity never reappeared in the CA1 pyramidal cells at 48-h reperfusion period. With DF2, the ubiquitin immunoreactivity at no time disappeared during 48 h of reperfusion period from the hippocampal neurons including CA1 pyramidal cells. *DG*, dentate granule (cells)

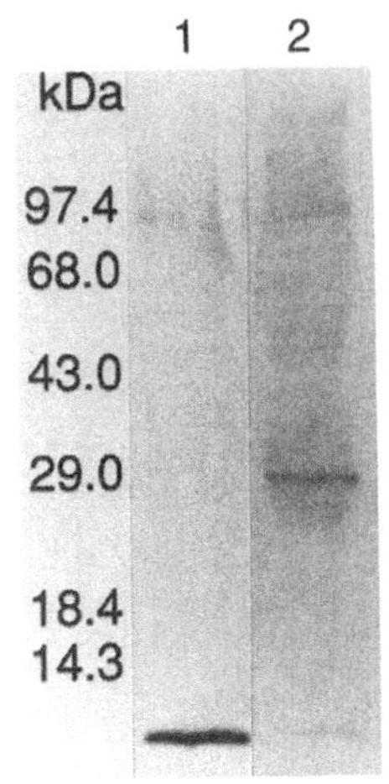

Fig. 2. Immunoprecipitation of supernatant from control gerbil brain homogenates with U-5379 (*lane 1*), and DF2 (*lane 2*). The immunoprecipitates were electrophoresed, transferred, and probed with DF2. U-5379 immunoprecipitated a large amount of free ubiquitin (*lane 1*), while DF2 precipitated only a trace amount of ubiquitin (*lane 2*). The high-molecular-weight regions in *lane 2* appeared to stain more strongly than that in *lane 1*

sistent with this, Hayashi et al. [2] showed an increase in Triton X-100-insoluble, high-molecular-weight conjugates in the CA1 after 20-min rat forebrain ischemia followed by 24 h of reperfusion.

Overall, transient ischemia depletes free ubiquitin, but not conjugated ubiquitin in the postischemic CA1. Since hippocampal neurons accumulate denatured proteins after brief forebrain ischemia in the gerbil [6], much more free ubiquitin is probably needed for degradation of abnormal proteins. However, since global protein synthesis is severely suppressed in the postischemic CA1 [7], the same may also be said for ubiquitin. If ubiquitinated proteins would be efficiently degraded, a significant amount of free ubiquitin would be released and recycled; immunostaining indicates that this is not the case. Therefore, the depletion of free ubiquitin causes in turn intracellular accumulation of abnormal proteins, leading to severe metabolic derangement and possibly contributing to neuronal cell death.

References

1. Bond U, Schlesinger MJ (1985) Ubiquitin is a heat shock protein in chicken embryo fibroblasts. Mol Cell Biol 5: 949–956
2. Hayashi T, Takada K, Matsuda M (1992) Subcellular distribution of ubiquitin-protein conjugates in the hippocampus following transient ischemia. J Neurosci Res 31: 561–564
3. Magnusson K, Wieloch T (1989) Impairment of protein ubiquitination may cause delayed neuronal death. Neurosci Lett 96: 264–270
4. Mori H, Kondo J, Ihara Y (1987) Ubiquitin is a component of paired helical filaments in Alzheimer's disease. Science 235: 1641–1644
5. Rechsteiner M, Hoffman L, Dubiel W (1993) The multicatalytic and 26S proteases. J Biol Chem 268: 6065–6068
6. Siman R, Noszek JC (1988) Excitatory amino acids activate calpain I and induce structural protein breakdown in vivo. Neuron 1: 279–287
7. Thilmann R, Xie Y, Kleihues P, Kiessling M (1986) Persistent inhibition of protein synthesis precedes delayed neuronal death in postischemic gerbil hippocampus. Acta Neuropathol (Berl) 71: 88–93

Insight into the Protective Mechanisms of Cortical Spreading Depression in Cardiac Arrest Cerebral Ischemia in Rat

N. KAWAHARA[1], C.A. RUETZLER[1], G. MIES[2], S.D. CROLL[3], S.J. WIEGAND[3], and I. KLATZO[1]

Summary

Cortical spreading depression has recently been shown to protect hippocampal CA1 pyramidal neurons, as well as cortical neurons, against ischemic damage if elicited prior to ischemia. The present study was undertaken to investigate pathomechanisms of neuronal protection conferred by cortical spreading depression (CSD) elicited by the topical application of 5 *M* KCl to the occipital cortex in Sprague-Dawley rats. Autoradiographic analysis on brain tissue metabolism revealed an increase of cerebral protein synthesis 3 days after CSD, whereas a reduction of glucose utilization was observed in both the CSD and control (NaCl application) groups. Immunohistochemistry disclosed an enhanced expression of c-Fos, glial fibrillary acidic protein (GFAP), and basic fibroblast growth factor after CSD, whereas heat-shock protein (hsp) 72 was not induced throughout an observation period of up to 7 days. Northern blot analysis for brain-derived neurotrophic factor (BDNF) revealed a biphasic increase in mRNA, the first peak appearing at 4 h and the second at 3 days after CSD. Quantitative measurement of BDNF by enzyme linked immunosorbent assay (ELISA) was in accordance with the biphasic pattern of mRNA, although the second peak lasted up to 7 days. These results show that neither metabolic suppression nor induction of hsp72 are related to postischemic neuronal protection following CSD, and suggest that stimulation of neurotrophic factors prior to an ischemic insult is a possible mechanism for neuronal resistance.

Introduction

Measures to ameliorate ischemic brain damage have been pursued for many years. Following the discovery that preconditioning by short sublethal cerebral ischemia can be protective against a subsequent lethal period of ischemia – a phenomenon known as "ischemic tolerance" [4, 5] – extensive studies have been centered on the elucidation of endogenous mechanisms for protection against ischemia. In a previous study, we demonstrated that cortical spreading depression (CSD) elicited 3 days prior to

[1] Stroke Branch, National Institutes of Health, Bldg 36, Rm4A-03, 36 Convent Drive, MSC 4128, Bethesda, MD 20892-4128, USA
[2] Max-Planck Institute for Neurological Research, Glueler Str. 50, 50931 Cologne, Germany
[3] Regeneron Pharmaceuticals Inc., 777 Old Saw Mill River Road, Tarrytown, NY 10591, USA

Maturation Phenomenon in Cerebral Ischemia II
U. Ito et al. (Eds.)

lethal ischemia could also induce a tolerant state in hippocampal CA1 pyramidal neurons [3]. This effect was not observed when CSD was induced 1 or 7 days prior to ischemia. Protection of cortical neurons was further confirmed 24h after CSD [6]. The purpose of the present study was to elucidate changes in brain tissue associated with transient neuronal protection following transient CSD, with special reference to alterations in brain metabolism and changes in expression of hsp72 and growth factors, such as basic fibroblast growth factor (bFGF) and brain-derived neurotrophic factor (BDNF).

Materials and Methods

Cortical spreading depression was elicited in Sprague-Dawley rats by topical application of 5*M* KCl to the exposed dura on the left occipital cortex for 1h as described previously [3]. Sham operated rats and rats in which KCl was substituted by NaCl of equal concentration served as controls.

Alterations in metabolism were assayed using autoradiographic analyses for glucose utilization and protein synthesis in rats sacrificed 3 days following CSD by use of a double tracer technique (^{14}C deoxyglucose and ^{3}H-leucine, $n = 4$).

Immunohistochemical evaluation of c-Fos, hsp72, glial fibrillary acidic protein (GFAP), and bFGF was carried out in rats sacrificed at 3h, and 1, 3 and 7 days following induction of CSD ($n = 4$ in each group), and brains were processed for immunohistochemical analysis using an antibody against c-Fos (Oncogene Science), hsp72 (Amersham), GFAP (Sigma), and bFGF (Oncogene Science).

In another group of rats the expression of BDNF mRNA was examined by Northern blot analysis and in situ hybridization. Cortical and hippocampal total RNA from each hemisphere was isolated at 4h and 24h, and 3 and 7 days following either KCl or NaCl application, and hybridized to a ^{32}P-labeled cDNA probe for BDNF. In in situ hybridization studies, fixed brain sections at the same time intervals were hybridized to a ^{35}S-labeled cDNA probe for BDNF and exposed to autoradiographic film for several days. For quantitative measurement of BDNF, its protein content was measured using two-site enzyme linked immunosorbent assay (ELISA) in cortical and hippocampal homogenates taken at 4h, 12h and 24h, and 3 and 7 days after electrolyte application ($n = 4$ in each group).

Results

Evaluation of glucose utilization and cerebral protein synthesis on the same section disclosed a marked reduction of glucose utilization in the ipsilateral hemisphere 3 days after KCl application, while cerebral protein synthesis was found to be discretely increased. A reduction of glucose utilization in the ipsilateral hemisphere was also observed following NaCl application, but cerebral protein synthesis remained unchanged (Fig. 1).

Immunohistochemical analysis for c-Fos exhibited conspicuously enhanced immunostaining of neurons throughout the ipsilateral cortex and bilaterally in the dentate gyri of the hippocampus only in rats subjected to CSD and sacrificed 3h after

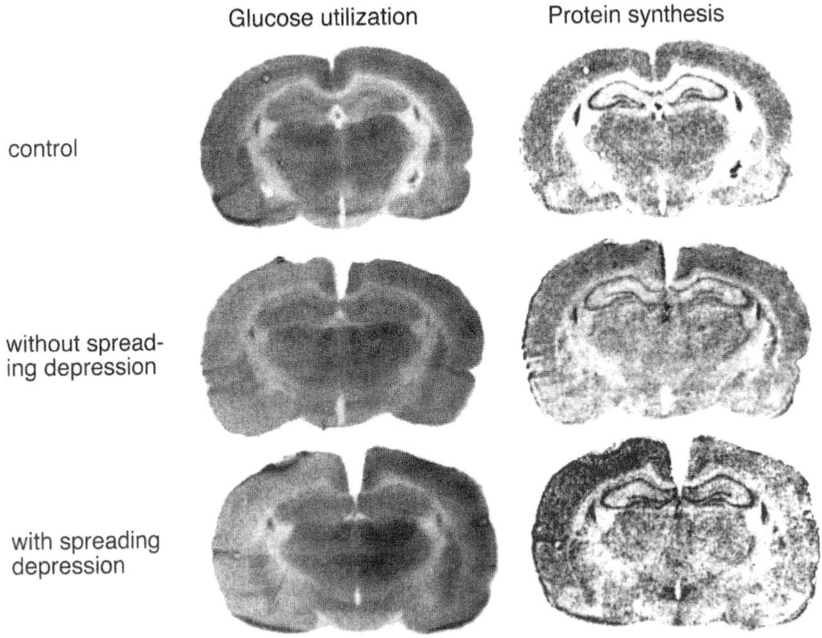

Fig. 1. Autoradiography of cerebral protein synthesis and glucose utilization in rats sacrificed 3 days after electrolyte application and in sham operated rats. Glucose utilization in the ipsilateral cortex and thalamus was markedly reduced in both NaCl- and KCl-treated rats, whereas an increase of protein synthesis in the ipsilateral cortex and lateral aspect of the hippocampal formation was observed only in KCl-treated animals

the application of KCl (Fig. 2). On the other hand, immunostaining for the inducible type of hsp72 was negative in all groups of rats at any time point after application of either KCl or NaCl. In contrast, GFAP immunostaining disclosed a marked enhancement of staining of astrocytes throughout the ipsilateral cortex, beginning 3 days after the application of KCl and lasting up to 7 days. In NaCl control rats, variably increased staining of astrocytes was noted mostly around the application site at 3 days, which later involved a wider region in the ipsilateral cortex. In the hippocampus, however, GFAP immunostaining did not show any appreciable changes compared to normal controls in either group. Immunostaining against bFGF revealed similar changes in glial cells. KCl application enhanced glial expression of bFGF in the entire ipsilateral cortex at 3 and 7 days, while NaCl application slowly increased the intensity of staining around the application site beginning at 3 days, with involvement of a wider region of cortex by 7 days.

Northern blot analysis demonstrated biphasic upregulation of BDNF mRNA in the ipsilateral cortex (Fig. 3). A first peak was detected at 4 h and a second peak at 3 days after application of KCl. In contrast, expression of mRNA in NaCl-treated rats did not reveal cortical changes. In the hippocampus, on the other hand, the level of mRNA remained unchanged up to 7 days in either KCl- or NaCl-treated rats. In situ hybridization studies confirmed the results of Northern blot. Specifically, the elevation was most marked in cortical layer II and VI in KCl-treated rats, while little change was noted in the hippocampus (data not shown).

BDNF assays of cortical homogenates revealed a biphasic increase in the ipsilateral cortex after KCl application which is in agreement with changes in mRNA of this neurotrophic factor (data not shown). The first peak appeared at 12h and the second was detected 3–7 days after CSD induction. In NaCl-treated rats, no such alterations in BDNF levels were observed. BDNF content in KCl-treated animals was significantly larger compared to that in NaCl-treated rats.

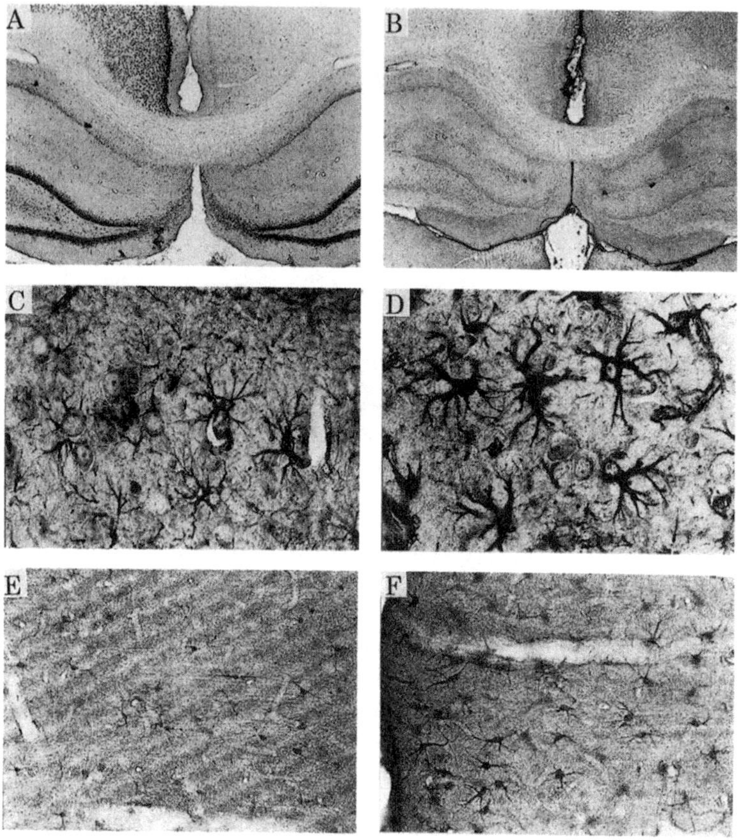

Fig. 2A–F. Immunohistochemistry in rats subjected to cortical spreading depression (CSD). c-Fos was broadly upregulated in the cortex on the ipsilateral side and in the dentate gyrus bilaterally 3h after CSD (**A**), whereas heat shock protein (hsp) 72 immunoreactivity was not detected at any time point (**B**, 3h after CSD). Glial fibrillary acidic protein (GFAP) immunostaining revealed marked enhancement of the intensity of staining in the ipsilateral cortex 3 days after CSD (**D**), while the contralateral side did not show any change (**C**). Astrocytes also demonstrated increased immunoreactivity for basic fibroblast growth factor (bFGF) 3 days after CSD in the ipsilateral cortex (**F**), as compared to reactivity in the contralateral, control side. Original magnification: **A** and **B**, ×20; **C** and **D**, ×400; **E** and **F**, ×100; **C** and **D**, counterstained with cresyl violet

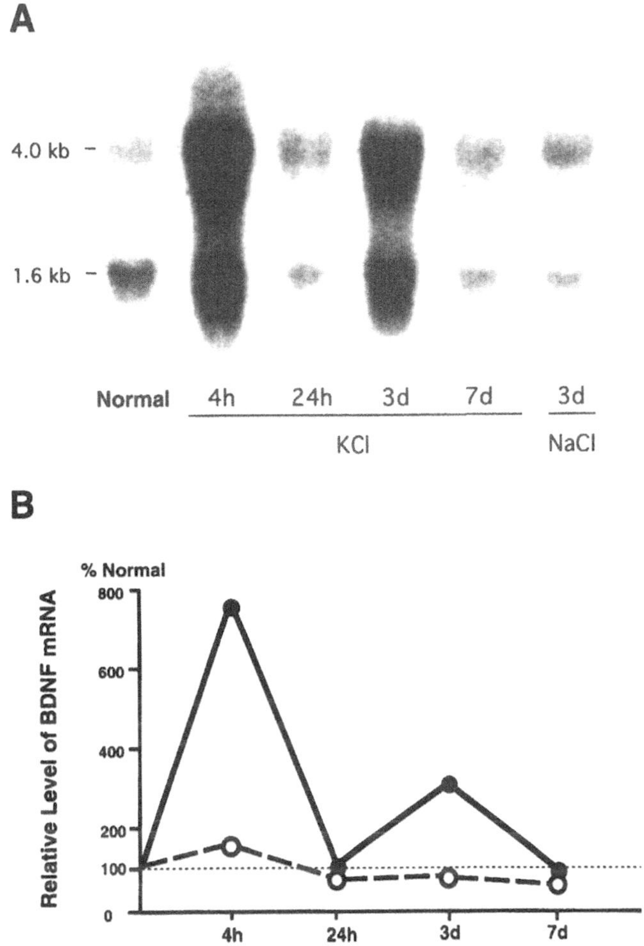

Fig. 3A,B. Northern blot analysis of brain-derived neurotrophic factor (BDNF) mRNA revealed a biphasic upregulation of both BDNF transcripts (4.0 and 1.6 kb) at 4 h and 3 days (**A**). Measurement of optical density (**B**) showed seven- and three-fold increases in the ipsilateral cortex (*solid line*) following cortical spreading depression (CSD), but not in NaCl-treated rats (*broken line*)

Discussion

At the tolerant period (3 days after CSD), glucose utilization was reduced in both KCl- and NaCl-treated rats, demonstrating that suppression of metabolic demand is not a major mechanism of protection following CSD. On the other hand, a general but discrete stimulation of protein synthesis in the ipsilateral hemisphere was observed only in KCl-treated rats. This finding led us to speculate that some protective substances might be produced at 3 days following CSD.

In accordance with this hypothesis, one of the transcription factors, c-Fos, was upregulated in the ipsilateral cortex and the dentate gyri of the hippocampus as analysed by immunohistochemical methods, suggesting a CSD-related alteration of gene expression. However, the putative protective protein hsp72 was not induced after CSD which is in line with previous findings [6], demonstrating that endogenous neuroprotective mechanisms are unrelated to hsp72. As has been reported previously [9], CSD markedly stimulated astrocytic cells as shown by GFAP immunostaining, which is also obvious 3 days after CSD in the ipsilateral cortex. This increase in GFAP levels was associated with enhanced staining of bFGF, the latter factor thought to be neuroprotective against an ischemic insult [8].

BDNF is a member of the neurotrophin family and has been shown to be upregulated by various insults and to be neuroprotective in vitro [2] as well as in vivo, in particular after recovery from an ischemic insult [1, 10]. The present study is the first to demonstrate that CSD elicits a biphasic increase in cortical mRNA which peaks at 4h and 3 days and coincides with the protective period, although an early peak in BDNF mRNA 4h after CSD has already been reported by Kokaia et al. [7]. BDNF protein levels also revealed a biphasic pattern, which was sustained up to 7 days. In contrast, there was no increase of BDNF mRNA in the hippocampus, although the protein transiently increased at 12h.

While the temporal and regional profiles of bFGF and BDNF did not match exactly with the protective period, i.e., 1 day in the cortex [6] and 3 days in the hippocampus [3], our results suggest involvement and stimulation of neurotrophic factors during the development of neuronal tolerance against ischemia. Further studies on the elucidation of the role of neurotrophic factors and their therapeutic application in cerebral ischemia is warranted.

References

1. Beck T, Lindholm D, Castrén E, Wree A (1994) Brain-derived neurotrophic factor protects against ischemic cell damage in rat hippocampus. J Cereb Blood Flow Metab 14: 689–692
2. Cheng B, Mattson MP (1994) NT-3 and BDNF protect CNS neurons against metabolic/excitotoxic insults. Brain Res 640: 56–67
3. Kawahara N, Ruetzler CA, Klatzo I (1995) Protective effect of spreading depression against neuronal damage following cardiac arrest cerebral ischaemia. Neurol Res 17(1): 9–16
4. Kirino T, Tsujita Y, Tamura A (1991) Induced tolerance to ischemia in gerbil hippocampal neurons. J Cereb Blood Flow Metab 11(2): 299–307
5. Kitagawa K, Matsumoto M, Kuwabara K, Tagaya M, Ohtsuki T, Hata R, Ueda H, Handa N, Kimura K, Kamada T (1991) "Ischemic tolerance" phenomenon detected in various brain regions. Brain Res 561(2): 203–211
6. Kobayashi S, Harris VA, Welsh FA (1995) Spreading depression induces tolerance of cortical neurons to ischemia in rat brain. J Cereb Blood Flow Metab 15(5): 721–727
7. Kokaia Z, Gidö G, Ringstedt T, Bengzon J, Kokaia M, Siesjö BK, Persson H, Lindvall O (1993) Rapid increase of BDNF mRNA levels in cortical neurons following spreading depression: regulation by glutamatergic mechanisms independent of seizure activity. Mol Brain Res 19: 277–286
8. Koketsu N, Berlove DJ, Moskowitz MA, Kowall NW, Caday CG, Finklestein SP (1994) Pretreatment with intraventricular basic fibroblast growth factor decreases infarct size following focal cerebral ischemia in rats. Ann Neurol 35: 451–457
9. Kraig RP, Dong L, Thisted R, Jaeger CB (1991) Spreading depression increases immunohistochemical staining of glial fibrillary acidic protein. J Neurosci 11: 2187–2198
10. Tsukahara T, Yonekawa Y, Tanaka K, Ohara O, Watanabe S, Kimura T, Nishijima T, Taniguchi T (1994) The role of brain-derived neurotrophic factor in transient forebrain ischemia in the rat brain. Neurosurgery 34: 323–331

Transneuronal Induction of Tolerance in Cerebral Ischemia

N. Kawahara[1], L. Belayev[2], F. Orzi[3], V. Colangelo[3], and I. Klatzo[3]

Summary

Development of enhanced tolerance to ischemic injury has been demonstrated when lethal ischemia is preceded by sublethal ischemic exposure. This study demonstrates a transneuronal induction of tolerance in distant areas, unaffected by primary ischemia, but connected by neuronal circuitry. In Model I, the induction of spreading depression in the cerebral cortex of the rat preceded 10-min cardiac arrest cerebral ischemia by 3 days. In Model II, the middle cerebral artery (MCA) occlusion for 2h in rats was followed 3 days later by 10-min clamping of both common carotid arteries, with simultaneous lowering of mean arterial blood pressure to 50 mm Hg. In Model III, 5-min bilateral occlusion of common carotid arteries in gerbils was preceded 3 days earlier by 35-min occlusion of the left common carotid artery in ischemia-sensitive animals. Our observations in Models I and II revealed a marked bilateral preservation of CA1 pyramidal neurons. Gerbils in Model III showed, on the side of the 35-min occlusion, numerous infarcted regions, including hippocampus, whereas the contralateral hippocampus revealed a marked preservation of CA1 neurons. It is assumed that CA1 pyramidal cell protection is transneuronally transmitted via the entorhinal cortex and is associated with early activation of c-fos expression in both hippocampi.

Introduction

It has been demonstrated that sublethal stress or injury may render neuronal tissue resistant to a subsequent lethal insult. Thus, Kitagawa et al. [7] and Kirino et al. [6] reported in gerbils that a preceding episode of brief ischemia can protect CA1 pyramidal neurons of the hippocampus against a lethal 5-min ischemic exposure. Enhanced neuronal resistance was similarly demonstrated after a brief (3-min) ischemic exposure in rats [11], as well as following oxidative stress [12], and in a chronic

[1] Department of Neurosurgery, University of Tokyo Hospital, 7-3-1 Hongo, Bunkyo-Ku, Tokyo 113, Japan
[2] Department of Neurology, University of Miami, School of Medicine, P.O. Box 016960, Miami, FL 33101, USA
[3] Istituto Scientifico Neurologico Mediterraneo Neuromed, Research Laboratories Loc. Camerelle, 86077 Pozzili, Italy

Maturation Phenomenon in Cerebral Ischemia II
U. Ito et al. (Eds.)

hypoperfusion model in gerbils [13]. In recent studies on spreading depression (SD), a statistically significant resistance to subsequent cardiac arrest cerebral ischemia (CACI) in rats was noted bilaterally in the CA1 sectors of the hippocampus, although DC recordings revealed SD confined only to the cortex of the hemisphere exposed to local application of KCl [4]. These observations strongly suggest a mechanism of transneuronal transmission of tolerance induction, since the hippocampus contralateral to SD was free from any direct impact of the ischemia. In the present study, in addition to the SD model, the phenomenon of transneuronal induction of tolerance was evaluated also in two other models in which the effect of unilateral ischemic injury was assessed with regard to its potential to induce neuronal tolerance demonstrable distantly in the contralateral hemisphere.

Material and Methods

In Model I, induction of SD in Sprague-Dawley rats preceded CACI by 3 days. The SD was initiated by the local application of 5 *M* KCl to exposed intact dura in the left occipital region. In order to assess the incidence of SD in various regions, the DC potentials were recorded from electrodes placed bilaterally in the cortex and left hippocampus. Three days after SD induction the rats were subjected to CACI which was produced by compression of the major cardiac vessels, according to the method previously described [4, 5]. In control animals the CACI was carried out without preceding SD induction. To evaluate an effect of SD alone on brain tissue, brains of rats sacrificed at various intervals were subjected to c-fos and brain-derived neurotrophic factor (BDNF) assays.

In Model II, Sprague-Dawley rats underwent a 2-h right middle cerebral artery occlusion (MCAo) by retrograde insertion of an intraluminal nylon suture coated with poly-L-lysine. Two-vessel occlusion (2Vo) of 10-min duration was carried out 3 days later by bilateral clamping of the common carotid arteries, with simultaneous lowering of mean arterial blood pressure (MABP) to 50 mm Hg. The control group consisted of animals which underwent sham MCAo and 2Vo. Rats were sacrificed 4 days after 2Vo, the brains fixed by transcardiac paraformaldehyde perfusion, sections stained by hematoxylin and eosin (H&E), and preserved neurons counted in the three CA1 subregions.

In Model III, 5-min bilateral common carotid artery occlusion (BCCAo) was preceded 3 days earlier by 35-min occlusion of the left common carotid artery (LCCAo) in gerbils which, by clinical signs, proved to be sensitive to unilateral carotid occlusion. Control animals had undergone 5-min BCCAo only. All animals were sacrificed 2 weeks after BCCAo and the preserved CA1 pyramidal neurons were counted bilaterally in cresyl violet-stained sections.

Results

In Model I, the DC recordings revealed typical recurrent SD waves only in the left cerebral cortex, whereas SD was absent in the cortex contralateral to KCl application, as well as in the left hippocampus. Brain sections of rats sacrificed 2 weeks after CACI had statistically significant enhancement of CA1 neuronal preservation on both sides, in comparison to rats which had undergone CACI only [4]. In animals sacrificed 3 h after SD, c-fos immunohistochemistry revealed markedly enhanced c-fos expression throughout the left cerebral cortex, particularly conspicuous in the entorhinal cortex, and, bilaterally, in the fascia dentata of the hippocampus (Fig. 1A).

In Model II, rats subjected to right MCAo followed by 2Vo showed cortical infarction on the side ipsilateral to MCAo, predominantly in the parietal region, with sparing of the hippocampus and entorhinal cortex (Fig. 2A). Neuronal cell counts in

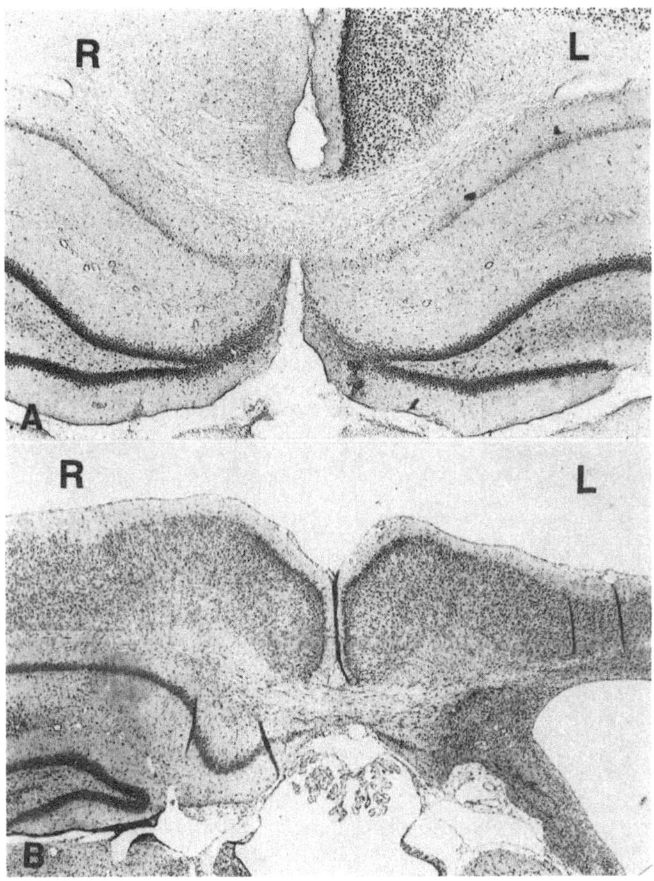

Fig. 1. **A** Model I: c-fos immunostaining in the rat subjected 3 hours earlier to spreading depression (SD) induction in the left cerebral cortex, revealing marked c-fos immunoreactivity, primarily in neurons. Strong c-fos positive staining is also shown bilaterally in the hippocampal dentate gyri, ×350; *R*, right, *L*, left. **B** Model III: gerbils subjected to 35-min left common carotid artery occlusion (LCCAo) and 5-min bilateral common carotid artery occlusion (BCCAo), sacrificed after 2 weeks. The left (*L*) hemisphere reveals infarction of the hippocampus, whereas the contralateral hippocampus, including CA1, appears fully preserved; Nissl stain, ×280; *R*, right; *L*, left

these animals revealed a significantly higher number of preserved CA1 neurons in both hippocampi than in CA1 sectors of control (sham MCAo + 2Vo) rats (Figs. 2A, B; Table 1).

In Model III, gerbils with 35-min LCCAo displayed areas of severe ischemic injury and infarction on the side of unilateral occlusion, observed mostly in the frontoparietal cortex, striatum, hippocampus and thalamus, whereas the entorhinal cortex appeared to be free of injury. The majority of gerbils in this group, in which 35-min LCCAo was followed 3 days later by 5-min bilateral occlusion, demonstrated severe destruction of the hippocampus on the side ipsilateral to LCCAo, whereas the contralateral hippocampus showed a marked preservation of CA1 pyramidal neurons (Fig. 1B), in comparison with control gerbils subjected to 5-min bilateral common carotid artery occlusion alone.

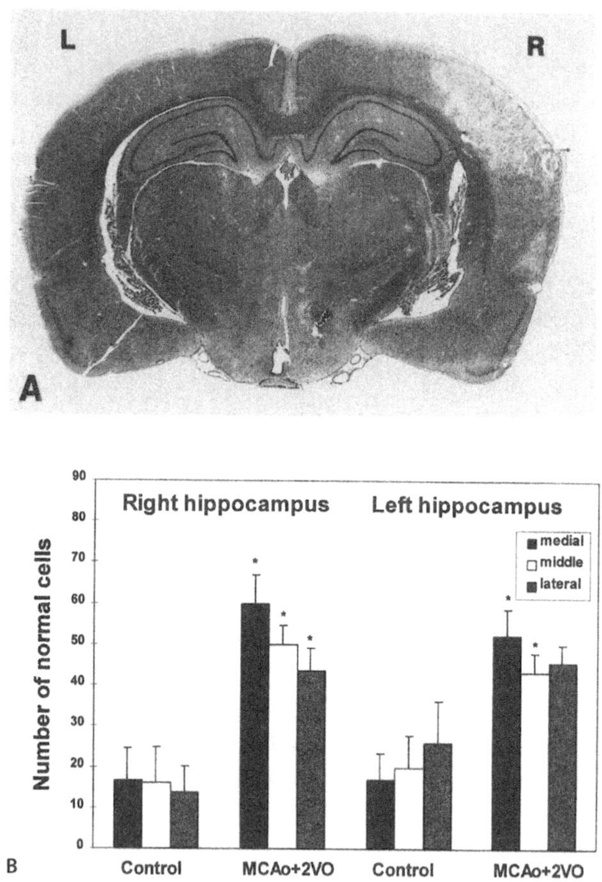

Fig. 2. **A** Model II: rat subjected to right middle cerebral artery occlusion (MCAo) for 2 h, followed 3 days later by 10-min two-vessel occlusion (2Vo). The right hemisphere reveals infarction in the parietal cortex while the entorhinal cortex (ERC) and hippocampi appear bilaterally well-preserved; hematoxylin and eosin (H&E) stain. **B** Distribution of neuronal damage in control (sham MCAo + 2Vo) and MCAo + 2Vo rats in the medial, middle and lateral regions of the CA1 sector of the hippocampus; H&E stain

Table 1. Histopathological outcome following middle cerebral artery occlusion (MCAo) + two-vessel occlusion (2Vo) in rats

Group	No.	Right side (MCAo) CA 1 sector of the hippocampus			Left side CA 1 sector of the hippocampus		
		Medial	Middle	Lateral	Medial	Middle	Lateral
MCAo + 2Vo	601291	36	36	22	51	52	49
	601292	68	43	48	26	23	24
	601302	87	67	51	56	43	45
	602061	62	49	41	72	56	51
	602063	59	60	63	61	46	57
	602121	47	45	36	48	40	47
Average		59.8	50.0	43.5	52.3	43.3	45.5
S.D.		17.6	11.5	14.0	15.4	11.6	11.3
S.E.		7.2	4.7	5.7	6.3	4.7	4.6
MCA sham + 2Vo	601162	4	3	2	2	2	2
	601293	7	3	7	19	21	34
	601303	35	45	31	26	38	47
	602162	36	28	27	35	37	46
	602202	1	1	2	3	1	1
Average		16.6	16.0	13.8	17.0	19.8	26.0
S.D.		17.4	19.7	14.1	14.4	18.0	22.9
S.E.		7.8	8.8	6.3	6.4	8.1	10.3
t-test (MCAo vs MCA sham)		0.003	0.006	0.007	0.004	0.03	0.1

Discussion

Observations from the three models presented have shown an induction of neuronal protection against lethal ischemic injury in areas distant from the territory of ischemic exposure. Since these distant regions were not directly affected by any ischemic disturbances in energy metabolism, a transneuronal mechanism of tolerance induction in these regions offers an acceptable explanation. In our studies, the entorhinal cortex (ERC) appears to serve as an important transneuronal relay point, since the ERC of each hemisphere provides, through the perforant pathway, innervation to the dentate gyri of both hippocampi. Thus, stimulation of the ERC on one side will transmit messages, including those related to tolerance induction, also to the contralateral hippocampus which originally may be untouched by ischemia. In support of the ERC's involvement is the demonstration of conspicuous, bilateral c-fos immunoreactivity in the dentate gyri (Fig. 1A), as well as the absence of any evident injury to the ERC in all three models examined.

The role of excitotoxic mechanisms in CA1 ischemic injury has been recognized as related to an excessive release of glutamate [1] which is activated by glutamatergic pathways originating in the ERC and then reaching the CA1 via the perforant pathway and Schaeffer's collaterals. Thus, it was demonstrated that transection of Schaeffer's collaterals results in an amelioration of CA1 neuronal damage [15]. Associated with glutamate release is the intraneuronal entry of calcium, which has been shown to activate c-fos and other early genes [2] in areas initially affected by ischemia, irrespective of whether neurons are destined to die or to recover. In areas of infarction the disintegration of cells becomes associated with the disappearance of c-fos immunoreactivity, whereas in the infarct-surrounding areas c-fos neuronal staining may persist for days [10, 14]. That elevated, transitory c-fos expression can be induced in stimulated but uninjured neurons is evident from our observations in Model 1, as well as from the study by Kokaia et al. [9], who described transitory elevation of c-fos expression in the hippocampus contralateral to MCAo. On the other hand, Liu and Chen [10] observed that neurons with persistent c-fos immunoreactivity undergo irreversible ischemic damage. Otherwise, the glutamatergic nature of c-fos induction is evident from its sensitivity to blocking by *N*-methyl-D-aspartate (NMDA) receptor antagonists, such as MKI-801 [3, 14].

The frequently described induction of tolerance in the CA1 sector of the hippocampus is not a unique phenomenon and is not limited only to this structure. Thus, Kitagawa et al. [8] demonstrated that if a brief (2-min) ischemia in gerbils is followed by longer (10-min) bilateral carotid occlusion, a protective effect of the first insult is recognizable also in the cerebral cortex, striatum and thalamus, as well as in the hippocampus.

The ability to create an enhanced tolerance and to influence recovery in various selected brain regions may have potential therapeutic significance. However, firstly, this would require elucidation of the precise neuronal circuitry involved in transneuronal induction. With such knowledge, it would be possible to stimulate topographically relevant points with an intensity which would not be harmful for brain tissue, but would be adequate to initiate protective mechanisms in selected regions. Secondly, the elucidation of the nature of various genetic events taking place would also be required and could narrow the search for the "magic bullets" truly responsible for the protection of cells against injury. Although the role of individual neurotrophic factors in tolerance remains obscure, c-fos appears to be a valuable marker for pointing out areas destined for the induction of tolerance.

References

1. Benveniste H, Drejer J, Schousboe A, Diemer NH (1984) Elevation of the extracellular concentrations of glutamate and aspartate in rat hippocampus during transient ischemia monitored by intracerebral microdialysis. J Neurochem 43: 1369–1374
2. Cole AJ, Saffen DW, Baraban JM, Worley PF (1989) Rapid increase of an immediate early gene messenger RNA in hippocampal neurons by synaptic NMDA activation. Nature 340: 474–476
3. Collaco-Moraes Y, Aspey BS, de Belleroche JS, Harrison MJ (1994) Focal ischemia causes an extensive induction of immediate early genes that are sensitive to MK-801. Stroke 25: 1855–1861
4. Kawahara N, Ruetzler CA, Klatzo I (1995) Protective effect of spreading depression against neuronal damage following cardiac arrest cerebral ischemia. Neurol Res 17: 9–16
5. Kawai K, Nitecka L, Ruetzler CA, Nagashima G, Joo F, Mies G, Nowak TS, Saito N, Lohr JM, Klatzo I (1992) Global cerebral ischemia associated with cardiac arrest in the rat: I. Dynamics of early neuronal changes. J Cereb Blood Flow Metab 12: 238–249
6. Kirino T, Tsujita Y, Tamura A (1991) Induced tolerance to ischemia in gerbil hippocampal neurons. J Cereb Blood Flow Metab 11: 299–307
7. Kitagawa K, Matsumoto M, Tagaya M, Hata R, Ueda H, Niinobe M, Handa N, Fukunaga R, Kimura K, Mikoshiba K, Kamada T (1990) "Ischemic tolerance" phenomenon found in the brain. Brain Res 528: 21–24

8. Kitagawa K, Matsumoto M, Kuwabara K, Tagaya M, Ohtsuki T, Hata R, Ueda H, Handa N, Kimura K, Kamada T (1991) "Ischemic tolerance" phenomenon detected in various brain regions. Brain Res 561: 203–211
9. Kokaia Z, Zhao Q, Kokaia M, Elmer E, Metsis M, Smith ML, Siesjö BK, Lindvall O (1995) Regulation of brain-derived neurotrophic factor gene expression after transient middle cerebral artery occlusion with and without brain damage. Exp Neurol 136: 73–88
10. Liu HM, Chen HH (1994) C-fos protein expression and ischemic changes in neurons vulnerable to ischemia/hypoxia, correlated with basic fibroblast growth factor immunoreactivity. J Neuropathol Exp Neurol 53: 598–605
11. Liu Y, Kato H, Nakata N, Kogure K (1992) Protection of rat hippocampus against ischemic neuronal damage by pretreatment with sublethal ischemia. Brain Res 586: 121–124
12. Ohtsuki T, Matsumoto M, Kawabara K, Kitagawa K, Suzuki K, Taniguchi N, Kamada T (1992) Influence of oxidative stress on induced tolerance to ischemia in gerbil hippocampal neurons. Brain Res 599: 246–252
13. Ohtsuki T, Matsumoto M, Kitagawa K, Taguchi A, Maeda Y, Hata R, Ogawa S, Ueda H, Handa N, Kamada T (1993) Induced resistance and susceptibility to cerebral ischemia in gerbil hippocampal neurons by prolonged but mild hypoperfusion. Brain Res 614: 279–284
14. Uemura Y, Kowall NW, Moskowitz MA (1991) Focal ischemia in rats causes time-dependent expression of c-fos protein immunoreactivity in widespread regions of ipsilateral cortex. Brain Res 552: 99–105
15. Wieloch T, Lindvall O, Blomqvist P, Gage FH (1985) Evidence of amelioration of ischemic neuronal damage in the hippocampal formation by lesions of the perforant path. Neurol Res 7: 24–26

III. Factors Modulating Neuronal Plasticity and Course of Maturation Phenomenon in Cerebral Ischemia

Features and Threshold of Infarct Development in Ischemic Maturation Phenomenon

U. Ito[1], S. Hanyu[2], Y. Hakamata[2], T. Kuroiwa[3], and M. Yoshida[2]

Summary

By repeating short ischemic insults, a single one of which did not cause significant neuronal loss, we developed a model to induce selective neuronal necrosis in the cerebral cortex without inducing focal infarction. Furthermore, we investigated the threshold for induction of focal infarction in the cerebral cortex by changing the number and interval between repetitive ischemic insults. We found that with an increase in number of these insults and a shortening of their interval, the ischemic injury intensified from less intensive to more intensive selective neuronal necrosis, and progressed to cerebral infarction. The maturation phenomenon of ischemic injury is expressed by a continuum from selective neuronal necrosis to infarction upon reaching a critical threshold of intensity of ischemic insults. Maturation of neuronal injury is associated with glycogen accumulation in the cortical tissue, but glycogen disappeared in infarcted areas of the cerebral cortex. The demonstrated selective graduation of different types of ischemic injury and their associated changes may be useful for the elucidation of mechanisms related to the induction of selective neuronal necrosis or of focal infarction, respectively.

Introduction

During the past decade investigations of ischemic brain injury have been focused mainly on selective neuronal necrosis rather than on the development of focal infarction. In view of the clinical significance of the latter, studies on developmental aspects, threshold values and therapeutic approaches for the prevention or amelioration of cerebral infarction have become urgently needed.

We have previously developed a model to induce selective neuronal necrosis in the cerebral cortex, by repeating short ischemic insults, a single one of which does not cause significant neuronal loss in the cortex [3]. By changing the number of repetitive insults, and the duration of intervening intervals, we have been able to establish a

[1] Department of Neurosurgery, Musashino Red Cross Hospital, 1-26-1, Kyonan-cho, Musashino-shi, Tokyo 180, Japan
[2] Department of Neurology, Jichi Medical School, 3311-1, Minami Kauchi-cho, Kauchi-gun, Tochigi Pref. 329-04, Japan
[3] Department of Neuropathology, Medical Research Center, Tokyo Medical and Dental University, 1-5-45, Yushima, Bunkyo-ku, Tokyo 113, Japan

Maturation Phenomenon in Cerebral Ischemia II
U. Ito et al. (Eds.)

threshold in the cerebral cortex for the progression of selective neuronal necrosis into the development of focal infarction [4, 7].

Materials and Methods

The left carotid artery of adult Mongolian gerbils was clipped under 2% halothane and 70% nitrous oxide and 30% oxygen anesthesia. Anesthesia was discontinued immediately after this procedure. Using the stroke index score [13], ischemia-positive animals were selected during the first 10-min occlusion. The experimental protocols are summarized in Table 1 and refer to the following animal groups: A, normal controls; B, single 10-min occlusion; C, occluded at 6-h intervals (subgroups a,b,c) with different numbers of additional 7-min occlusions; D, occluded at 12-h intervals (subgroups a,b,c) with different numbers of 7-min occlusions; in subgroups d and e, the last 7-min occlusion was carried out following 48-h or 96-h intervals, respectively; E, occluded at 24-h intervals with three additional 7-min ischemic periods; F, 48-h intervals following exposure to three additional 7-min occlusions. All animals were sacrificed 4 weeks after the start of ischemia, by transcardiac perfusion with 10% phosphate buffered formalin, except for group G which served to evaluate the temporal profile of neuronal survival, and in which 24-h interval animals were sarificed at 4, 7, and 28 days after the start of the ischemic procedure, respectively. The fixed brains were cut coronally at the level of the optic chiasma (face A) and infundibulum (face B). Paraffin sections of face A were stained with hematoxylin and eosin (H&E) and cresyl violet. The number of neurons in the area of ischemic cortex and in the symmetrically

Table 1. Time schedule of repeated ischemia and results

Repetition interval	One 10-min ischemia followed by 7-min ischemia with various repetition intervals	Group and sub-group	Percentage ratio of neuronal number (average ± SD)	Animals with infarction (*n*/total)
Control		A	108.8 ± 4.3	0/5
10-min single	10	B	101.3 ± 4.3	0/5
6-h interval	10 7	C a	$83.4 \pm 20.0^{*a}$	0/5
	10 7 7	b	$32.6 \pm 8.6^{*b}$	0/5 (0%)[1]
	10 7 7 7	c		4/5 (80%)[2]
12-h interval	10 7 7 7	D a	$63.4 \pm 20.1^{*c}$	0/5 (0%)[3]
	10 7 7 7 7	b		4/5 (80%)[4]
	10 7 7 7 7 7	c		5/5 (all cortical layers)
	10 7 7 7 48h 7	d		2/7 (29%)[5]
	10 7 7 7 96h 7	e		0/6 (0%)[6]
24-h interval	10 7 7 7	E	$72.3 \pm 5.3^{*d}$	0/5
48-h interval	10 7 7 7	F	$98.5 \pm 10.7^{*e}$	0/5

* Control, $p < 0.01$; difference among a, b, c, d, e $p < 0.05$ (ANOVA), 1–2, 3–4, 4–6 $p < 0.05$ (χ^2).

located area of non-ischemic cortex was counted as described elsewhere [3]. After adjustment for overestimation of the number of neurons in the left ischemic cortex due to postischemic hemispheric atrophy [3], the percentage ratio of the number of neurons in the left ischemic hemisphere to that in the right non-ischemic hemisphere was calculated.

In addition, in order to study the period when selective neuronal necrosis progresses to the development of focal infarction in group H, we have subjected gerbils to two 10-min episodes of ischemia at an interval of 5h. Brains were fixed and processed as described above at 15min, 5, 12, 24, 48 and 72h after the last ischemic insult. Paraffin sections were stained with H&E, cresyl violet and PAS. This model is designed to induce selective neuronal necrosis without infarction in the coronal sections cut at the level of infundibulum (face B), and to induce focal infarction of the third cortical layer in the more anteriorly situated coronal section cut at the level of the chiasma (face A).

Results

No significant neuronal loss was found following a single 10-min episode of ischemia (group B). In animals which showed no infarction, the average percentage ratio of preserved nerons in the cortex on the side of carotid occlusion was 63.4 ± 20.1, 72.3 ± 5.3 and 98.5 ± 10.7 in 12-, 24- or 48-h interval animals (group D-a, E, F), respectively. In group G, the average percentage ratio of surving neurons was 87.4 ± 20.0, 74.2 ± 6.2 and 72.3 ± 5.3, respectively, at 4, 7, and 28 days (Fig. 1B) after the start of ischemic procedure, indicating progression of neuronal loss during the first week.

In 6-h interval animals (group C), focal infarction in the third cortical layer was found in four of five gerbils which underwent three additional 7-min occlusions (subgroup c), whereas no infarction was found and the average percentage ratio of surviving neurons was 32.6 ± 8.6 and 83.4 ± 20.0 in animals with two or one additional 7-min ischemias (subgroup b,a), respectively.

In the 12-h interval group (D), focal infarction in the third cortical layer was found in four of five animals with four repetitive occlusions (subgroup b). The number of gerbils with focal infarction was reduced respectively to two of seven or none of six when the last interval was prolonged to 48h or 96h (subgroup d,e). All five animals showed large focal infarctions involving the entire cortical layer when additional 7-min occlusions were increased from four to five (subgroup c).

In group H, 15min after the last ischemia, no histological changes were recognized in the cerebral cortex, either in face B (selective neuronal necrosis later develops) or in face A (focal infarction later develops) sections. At a 5-h interval after the last ischemia, scattered necrotic neurons were found (Fig. 1A), and glycogen granules appeared diffusely (Fig. 1C) in all cortical layers, both in face B and A, especially in the third and fifth layers. At 12–24h, the number of necrotic neurons increased and glycogen granules slightly decreased in the cerebral cortex in face B. In face A, focal infarction became prominent in the third cortical layer, while glycogen granules disapeared. Proliferating capillaries, foamy macrophages and gemistocytic astrocytes

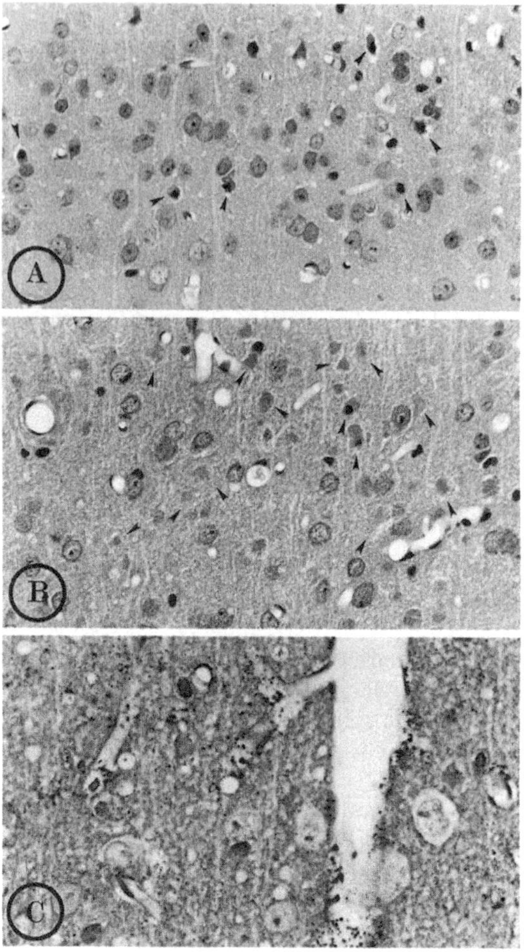

Fig. 1. A,B Selective necrosis of neurons in the third cortical layer. **A** At a 5-h interval after the last ischemia in group H where animals are subjected to two 10-min episodes of ischemia at an interval of 5 h, scattered necrotic neurons with dark cell body and pyknotic nucleus are found in the third cortical layer of face B (*arrows*). H&E, ×200. **B** At a 28-day interval after the start of the ischemic procedure in group E where unilateral carotid artery is occluded at 24-h intervals with three additional 7-min ischemic periods after 10-min initial occlusion, ghost neuronal cells with eosinophilic cell body and occasional pyknotic nucleus are found in the third cortical layer of face A (*arrows*). H&E, ×200. **C** Glycogen granules in the third cortical layer. At a 5-h interval after the last ischemia in group H where animals are subjected to two 10-min episodes of ischemia at an interval of 5 h, glycogen granules appear diffusely in third cortical layer of face B, especially prominently around the blood vessels. PAS, ×400

were seen in the infarcted area of face A from 48 to 72 h after the last ischemic insult. In face B, dead shrunken neurons with eosinophilic cytoplasm were found scattered among surviving neurons, whereas microglia and astrocytes were only slightly increased in number (Fig. 1B).

Discussion

In a single ischemic insult, the transition from selective neuronal necrosis to focal infarction in the cerebral cortex shows considerable variability [16]. However, by supplementing a single ischemic insult with repeated small ones we have achieved a more consistent and reproducible transition from selective neuronal necrosis to infarction. Using short ischemic insults, a single one of which does not cause significant cortical neuronal necrosis, we have developed an ischemia model whereby selective neuronal necrosis of varying intensity or focal infarction can be studied separately, according to the dosimetry of the repetitive ischemic insults.

A single 31-min ischemic insult produces a large focal infarction involving all cortical layers [5, 6]. However, when ischemia of 31 min is given divided into four insults of 10-, 7-, 7-, and 7-min occlusions, longer intervals between insults produce less intensive ischemic injury, e.g., at a 12-h interval, only neuronal necrosis without infarction; at a 24-h interval, still less intensive neuronal necrosis; at a 48-h interval, no significant neuronal necrosis can be observed (Table 1: Group D, E, F) [3, 8, 19].

By shortening the three intervals to 6 h, four out of five animals showed focal infarction, mainly in the third cortical layer (Table 1: Group C-c). By reducing the ischemia to two 7-min occlusions, only selective neuronal necrosis occurred, with 32.6% of neurons surviving (Group C-b). Further reduction to 10- and 7-min occlusions, separated by 6 h, resulted in an increase of surviving neurons to 83.4% (Group C-a). This suggests a threshold for transition from selective neuronal necrosis to infarction, between two and three repetitive 7-min ischemias at an interval of 6-h. In the 12-h interval group, no focal infarction was induced by three 7-min insults, and 63.4% of the neurons survived (Group D-a). By adding one additional episode of 7-min ischemia (namely, a fourth episode), four out of five animals showed focal infarction in the third cortical layer (Group D-b). Thus there is an apparent threshold for the transition to infarction occurring between three and four repetitions of the 7-min ischemic episodes [4]. By increasing the last interval of the fourth repetition from 12 to 48 h, infarction is observed in two out of seven animals (Group D-d): by increasing the last interval to 96 h, no infarction is observed (Group D-e). Also in this group, by adding a fifth episode of 7-min ischemia at an interval of 12 h, all animals demonstrated a large focal infarction (Group D-c) [7].

From the above findings, we can deduce that with an increase in frequency of insults and a decrease in duration of intervals, the ischemic injury progresses from occasional selective neuronal necrosis, to more neuronal necrosis, and, finally, to cerebral infarction. There must be a definite threshold level of ischemic insult to induce cerebral infarction [1]. Each small ischemic insult seems to induce some metabolic disarrangements in the cortical tissue that gradually normalize by reperfusion ([9, 10]; Kempski and Volk, this volume). When repetetive insults increase in number or the intervals become shorter, these biological disarrangements seem to become cumulative and to intensify the tissue injury [3, 4, 8, 12, 14, 15, 19].

The maturation phenomenon in cerebral ischemia has been previously defined as showing a delayed appearance and development after resumption of cerebral blood flow. The severity and rate of progression of the postischemic injury are directly related to the intensity of the ischemia, a less intense ischemic insult resulting in lesser and slower development of the lesions [5, 6]. From the present findings, we can assume that the maturation phenomenon is a continuous process from selective neuronal necrosis to infarction, depending on the amount of ischemic insult, and that there is a definite threshold level at which a transition from selective neuronal necrosis to infarction occurs.

It is generally accepted that glycogen in the brain is primarily confined to astrocytes and that it is constantly turning over, serving as a buffer for temporary imbalances between glucose utilization and glucose resupply from the cerebral circulation [17]. In the present study, the glycogen granules might be increased as a reaction to perturbed glucose metabolism after reperfusion [18] and/or to astrocytic gluconeogenesis which consumes lactic acid formed in the brain during anaerobic glycolysis [2]. Histologically, glycogen granules increased in the morphologically intact cortical tissue where infarction will develop at a later time. This finding suggests that the maturation phenomenon is still progressing in the brain tissue before necrosis begins [11].

Conclusions

Firstly, with an increase in the number of repetitive insults and shortening of the intervals, ischemic injury progresses from less extensive to more extensive neuronal necrosis, and then to cerebral infarction. Secondly, the maturation phenomenon is expressed by a continuum from selective neuronal necrosis to infarction upon reaching a critical threshold of intensity. Thirdly, maturation of neuronal injury is associated with glycogen accumulation. Finally, the present model is useful for studying various parameters of ischemic injury operating in the cerebral cortical tissue which are associated with the transition from selective neuronal necrosis to focal infarction.

References

1. DeGirolami U, Crowell RM, Marcoux FW (1984) Selective necrosis and total necrosis in focal cerebral ischemia. Neuropathologic observations on experimental middle cerebral artery occlusion in the macaque monkey. J Neuropathol Exp Neurol 43(1): 57–71
2. Dringen R, Schmoll D, Cesar M, Hamprecht B (1993) Incorporation of radioactivity from [14C]lactate into the glycogen of cultured mouse astroglial cells. Evidence for gluconeogenesis in brain cells. Biol Chem Hoppe Seyler 374(5): 343–347
3. Hanyu S, Ito U, Hakamata Y, Yoshida M (1993) Repeated unilateral carotid occlusion in Mongolian gerbils: quantitative analysis of cortical neuronal loss. Acta Neuropathol (Berl) 86(1): 16–20
4. Hanyu S, Ito U, Hakamata Y, Yoshida M (1995) Transition from ischemic neuronal necrosis to infarction in repeated ischemia. Brain Res 686: 44–48
5. Ito U, Spatz M, Walker J Jr, Klatzo I (1975) Experimental cerebral ischemia in mongolian gerbils. I. Light microscopic observations. Acta Neuropathol (Berl) 32(3): 209–223
6. Ito U, Yamaguchi T, Tomita H, Tone O, Shishido T, Hayashi H, Yoshida M (1992) Maturation phenomenon of ischemic injuries observed in Mongolian gerbils: introductory remarks. In: Ito U, Kirino T, Kuroiwa T, Klatzo I (eds) Maturation phenomenon in cerebral ischemia. Springer, Berlin Heidelberg New York, pp 1–13
7. Ito U, Hanyu S, Hakamata Y, Yoshida M (1995) Selective neuronal death and focal infarction in repeated ischemia. J Cereb Blood Flow Metab 15 [Suppl 1]: S199
8. Kato H, Kogure K (1990) Neuronal damage following non-lethal but repeated cerebral ischemia in the gerbil. Acta Neuropathol (Berl) 79(5): 494–500

9. Laptook AR, Corbett RJ, Ruley J, Olivares E (1992) Blood flow and metabolism during and after repeated partial brain ischemia in neonatal piglets. Stroke 23(3): 380–387
10. Matsumoto T, Obrenovitch TP, Parkinson NA, Symon L (1990) Cortical activity, ionic homeostasis, and acidosis during rat brain repetitive ischemia. Stroke 21(8): 1192–1198
11. Mrsulja BB, Mrsulja BJ, Ito U, Walker J Jr, Spatz M, Klatzo I (1975) Experimental cerebral ischemia in Mongolian gerbils. II. Changes in carbohydrates. Acta Neuropathol (Berl) 33(2): 91–103
12. Nowak T Jr, Tomida S, Pluta R, Xu S, Kozuka M, Vass K, Wagner HG, Klatzo I (1990) Cumulative effect of repeated ischemia on brain edema in the gerbil. Biochemical and physiological correlates of repeated ischemic insults. Adv Neurol 52: 1–9
13. Ohno K, Ito U, Inaba Y (1984) Regional cerebral blood flow and stroke index after left carotid artery ligation in the conscious gerbil. Brain Res 297(1): 151–157
14. Plum F (1983) What causes infarction in ischemic brain?: The Robert Wartenberg Lecture. Neurology 33(2): 222–233
15. Siesjo BK (1992) Pathophysiology and treatment of focal cerebral ischemia, part II: mechanisms of damage and treatment. J Neurosurg 77(3): 337–354
16. Smith ML, Auer RN, Siesjo BK (1984) The density and distribution of ischemic brain injury in the rat following 2–10 min of forebrain ischemia. Acta Neuropathol (Berl) 64(4): 319–332
17. Sokoloff L (1992) Energy metabolism and effects of energy depletion or exposure to glutamate. Can J Physiol Pharmacol 70(12): S107–S112
18. Swanson RA, Choi DW (1993) Glial glycogen stores affect neuronal survival during glucose deprivation in vitro. J Cereb Blood Flow Metab 13(1): 162–169
19. Tomida S, Nowak T Jr, Vass K, Lohr JM, Klatzo I (1987) Experimental model for repetitive ischemic attacks in the gerbil: the cumulative effect of repeated ischemic insults. J Cereb Blood Flow Metab 7(6): 773–782

Impending Cerebral Infarction and Inflammatory Response: A Possible Target for Therapeutic Intervention

T. Kogure[1], Y. Yamasaki[2], Y. Matsuo[2], and K. Kogure[2]

Summary

Energy failure due to decreased blood flow induces various changes in the structural and functional integrity of the brain cells. After a certain degree of ischemia, the changes become irreversible and the tissues undergo pan-necrosis. This activates a cascade of inflammatory response which scavenges the damaged cells in areas of impending infarction through liquification and phagocytosis. Induction of the immune response further accelerates disintegration of the damaged cells. Recent reports suggest that this inflammatory response may over-kill brain cells, enhancing the area of damage caused by the primary ischemic insult. Anti-inflammatory agents have been found to effectively decrease the size of the infarct in animal models of ischemia. Further research of inflammatory response in the ischemic brain may yield a new target for the therapy of cerebral ischemia.

Introduction

Innumerable experiments attempting to understand the development of cerebral infarctions have been performed by researchers in various fields of interest. In general, ischemic neuronal death has been understood as irreversible neuronal damage caused by energy failure. Depending on the size and duration of the primary ischemic insult, the irreversible damage leading to pan-necrosis is believed to develop within minutes to several days. Secondary ischemic damage induced during this period may enhance the size of the infarction. Recent experiments suggest that inflammatory response within the brain may promote ischemic cell injury and worsen the outcome. This finding may propose a new target for a potentially effective clinical method of therapeutic intervention against the development of cerebral infarction.

[1] Tokyo Jikei University School of Medicine, Department of Neurosurgery, 3-25-8 Nishi-Shinbashi, Minato-ku, Tokyo, Japan
[2] Foundation for Brain Function and Diseases, Institute of Neuropathology, 1216, Oaza Nakase, Fukaya-city, Saitama Prefect. 366, Japan

Maturation Phenomenon in Cerebral Ischemia II
U. Ito et al. (Eds.)

Early Ischemic Events

An ischemic insult is known to activate a cascade of molecular and biochemical changes. The primary event seen directly after ischemia is energy failure due to the rapid depletion of the high energy phosphate compound, adenosine triphosphate (ATP). This is caused by the insufficient supply of oxygen and glucose needed for the supplementation of ATPs. Consequently, functions within the cell that are ATP-dependent are inhibited causing various disturbances in cellular metabolism and ionic homeostasis.

The cessation of the ion pumps such as the Na^+-K^+ adenosine triphosphatase (ATPase) results in the accumulation of extracellular potassium ions and intracellular sodium ions causing membrane depolarization. This activates the voltage-dependent calcium channels and causes a rapid influx of Ca^{2+} [14].

Intracellular accumulation of Ca^{2+} in the presynaptic neurons activates the nonspecific, non-physiological release of excitatory amino acids (EAAs) such as glutamate from the presynaptic terminals [3]. This stimulates various metabotropic receptors which activates phospholipase C (PLC).

PLC mediates the hydrolysis of phophatidylinositol 4,5-diphosphate (PIP_2) and produces 1,2-diacylglycerol (DG) and inositol 1,4,5-triphosphate (IP_3). Furthermore, the decomposition of DG by di- and mono-acylglycerol lipases, and IP_3 stimulation upon receptors on the endoplasmic reticulum enhances the release of Ca^{2+} from intracellular stores, causing higher concentrations of intracellular Ca^{2+} [12]. EAAs may also stimulate receptors on the post synaptic neurons and enhance calcium influx by activating the Ca^{2+} channels.

The increase in the cytosolic Ca^{2+} causes nonspecific activation of various enzymes such as proteases, lipases, and protein kinases. These enzymes accelerate the decomposition of membranous structural phopholipids such as phosphatidyl choline (PC) and phosphatidyl ethanolamine (PE), causing degenerative changes in the structural integrity of the cell.

Post-ischemic Events

Obviously, brain tissue cannot survive a highly persistent ischemia where energy failure prevents physiological reconstitution of structural integrity and biological homeostasis. After a certain time point, the ischemic changes becomes irreversible and a large portion of the tissue undergoes infarction. However, effective reperfusion to the ischemic tissues before the point of no return may activate a restoration program to aid in salvaging cells within the ischemic region.

General protein synthesis in the ischemic brain cells is markedly suppressed [16] and a variety of genes, such as the immediate early genes (IEGs), are induced through phosphorylation of regulatory enzymes [2, 7]. Stress stimulations other than ischemia are also known to induce such changes in the genomic properties of the cells [10].

IEGs such as the fos families (c-fos, fosB) and the jun families (c-jun, junB, junD) act as messengers and encode proteins that contribute to the formation of various transcription factors. The fos/jun families contain the leucine zipper domain which forms activator protein-1 (AP-1) to bind to the tissue plasminogen activator (TPA) responsive element (TRE) in target DNAs.

These transcription factors are believed to play an important role in inducing the expression of various heat shock proteins (HSPs) [10] and other types of stress induced proteins such as the amyloid precursor proteins (APPs) [1], and superoxide dismutase (SOD), thus mediating a long term response to ischemia.

HSPs of different molecular weights, ranging from the 8-kDa ubiquitin to the 110-kDa HSP, are induced through various types of stress stimulation. These act as molecular chaperones, stabilizing the conformation of both normal and denatured proteins to acquire proper functional integrity. A stress protein especially induced by ischemic stress is HSP70, a protein with a molecular weight of 70kDa [6, 9].

This cascade of stress response (the expression of IEGs, induction of the corresponding transcription factors, and selective synthesis of stress proteins) may help the affected brain cells to survive the provoked ischemic changes and acquire tolerance to secondary ischemia [5, 8].

If the primary ischemic insult is severe and dysfunction of the cellular integrity exceeds the aforementioned stress-survival response, a variety of trophic factors which activate astrocytes and microglias within the vicinity are released. Astrocytes are known to normally produce glutamine synthetase, preventing excitotoxic neuronal damage caused by excessive release of glutamate in the synaptic space. Activation of astrocytes may increase the production of glutamine synthetase and help protect still-to-recover neurons within the ischemic region [13].

Inflammatory Response and Reperfusion Injury

Irreversible damage in the peripheral tissues leading to necrosis activates various inflammatory responses. These responses scavenge the necrotic tissues and allows proliferation of the surviving cells and affords aid in the reacquisition of structural and functional integrity.

A similar inflammatory response can be seen in the brain. Cytokines such as interleukin-1 (IL-1) interleukin-8 (IL-8) and tumor necrosis factor-α (TNF-α) facilitate the proliferation of astroglias and the induction of intercellular adhesion molecule-1 (ICMA-1) on the endothelial surface. Activated T-cells secrete TNF-α and interferon-γ (IFN-γ) and induce the major histocompatibility complex (MHC) class-I and class-II on the surface of neurons and astroglia [4]. Induction of these molecules is important in the endothelial adhesion, activation, and emigration of blood-borne inflammatory cells such as the leukocytes.

What makes the inflammatory response within the brain different from that of the peripheral tissues is that the neurons within the brain are unable to proliferate. In contrast to the peripheral tissues where inflammation is part of the healing process, inflammatory response within the brain may not be a beneficial activity for the cells surrounding the infarction. The cytotoxic agents released by the activated neutrophils and phagocytosis may injure cells in the vicinity of the infarction which are still recovering from the ischemic insult. This may increase the damage induced by the ischemic insult and the size of the infarct.

Reperfusion may be highly effective in preventing the expansion of ischemic injury, but after a certain time point reperfusion may introduce highly activated neutrophils and phagocytes to the ischemic focus as well as various reactive oxygen species and increase ischemic damage. However, the identification of this certain time point is still to be determined. Since induction of cytokines depend on DNA transcription and RNA translation, it is presumed that this critical time point exists at an early stage of ischemia.

New Targets for Therapeutic Intervention Against Cerebral Infarction

The inflammatory cascade consists of the release of cytokines, activation of neutrophils, induction of adhesion molecules, and chemoattraction of neutrophils into the damaged tissues. Each of these phases represents a possible target for therapeutic intervention.

Anti-neutrophil monoclonal antibodies (RP-3) administered to animal models of ischemia depleted or inactivated the leukocytes and minimized the size of the infarction [15]. The development of post-ischemic cerebral edema was also suppressed. X-ray irradiation [17] and colchicine administration [15] also exhibited similar effects.

Administration of anti-adhesion molecule antibodies such as the anti-ICAM-1 antibody (1A-29), anti-CD11 antibody (WT-1), and anti-CD18 antibody (WT-3), were also effective in suppressing post-ischemic brain injury [11].

Likewise, cytokine inhibiting agents such as PAF antagonist (Y-24810) (Matsuo et al., unpublished data), IL-1 blocking agent (ZnPP), anti-IL-1 antibody, and IL-1 release inhibitor also prevented the expansion of post ischemic brain injury [17].

Immunosuppressants such as cyclosporin A which attenuates exaggerated immunoreactions were also found to be effective in minimizing post-ischemic damage [15].

Further understanding of the inflammatory response in the brain may help in the integration of a highly effective clinical method of therapy for ischemic stroke.

References

1. Abe K, Tanzi RE, Kogure K (1991) Selective induction of Kunitz type protein inhibitor domain containing amyloid precursor protein mRNA after persistent focal ischemia in rat cerebral cortex. Neurosci Lett 125: 172–174
2. An G, Lin TN, Liu JS, Xue JJ, He YY, Hsu CY (1993) Expression of c-fos and c-jun family genes after focal cerebral is cerebral ischemia. Ann Neurol 33: 457–464
3. Benveniste H, Drejer J, Schousboe A, Diemer NH (1984) Elevation of the extracellular concentrations of glutamate and asparate in rat hippocampus during transient cerebral ischemia monitored by intracerebral microdialysis. J Neurochem 43: 1369–1374
4. Benveniste H (1992) Inflammatory cytokines within the central nervous system: sources, function and mechanism of action. Am J Physiol 263: C1–16
5. Kato H, Liu Y, Araki T, Kogure K (1991) Temporal profile of the effects of pretreatment with brief cerebral ischemia on the neuronal damage following secondary ischemic insult in the gerbil: cumulative damage and protective effects. Brain Res 553: 238–242

6. Kinouchi H, Sharp FR, Hill MP, Koistinaho J, Sagar SM, Chan PH (1993) Induction of 70-kDa heat shock protein and hsp70 mRNA following transient focal cerebral ischemia in the rat. J Cereb Blood Flow Metab 13: 105–115
7. Kinouchi H, Sharp FR, Chan PH, Koistinaho J, Sagar SM, Yoshimoto T (1994) Induction of c-fos, junB, c-jun, and hsp70 mRNA in cortex, thalamus, basal ganglia, and hippocampus following middle cerebral artery occlusion. J Cereb Blood Flow Metab 14: 808–817
8. Kitagawa K, Matsumoto M, Tagaya M, Hata R, Ueda H, Niinobe M, Handa N, Fukunaga R, Kimura K, Mikoshiba K, Kamada T (1990) Ischemic tolerance phenomenon found in the brain. Brain Res 528: 21–24
9. Li Y, Chopp M, Zhang ZG, Zhang RL, Garcia H (1993) Neuronal survival is associated with 72kDa heat shock protein expression after transient middle cerebral artery occlusion in the rat. J Neurol Sci 120: 187–194
10. Lindquist S, Craig EA (1994) The heat shock protein. Annu Rev Genet 22: 631–677
11. Matsuo Y, Onodera H, Shiga Y, Shozuhara H, Ninomiya M, Kihara T, Tamatani T, Miyasaka M, Kogure K (1994) Role of cell adhesion molecules in brain injury after transient middle cerebral artery occlusion in the rat. Brain Res 656: 344–352
12. Mitani A, Yanase H, Sakai K, Wake Y, Kataoka K (1993) Origin of intracellular Ca^{2+} elevation induced by in vitro ischemia like condition in hippocampal slices. Brain Res 601: 103–110
13. Petito CK, Chung MC, Verkhovsky LM, Coopwe AJL (1992) Brain glutamine synthetase increases following cerebral ischemia in the rat. Brain Res 569: 275–280
14. Siesjo BK, Bengtsson F (1989) Calcium fluxes, calcium antagonists, and calcium related pathology in brain ischemia, hypoglycemia, and spreading depression: a unifying hypothesis. J Cereb Blood Flow Metab 9: 127–140
15. Shiga Y, Onodera H, Matsumoto Y, Matsuo Y, Kogure K (1992) Cyclosporin A protects against ischemia reperfusion injury in the rat brain. Brain Res 595: 145–148
16. Thilmann R, Xie Y, Klehues P, Kiessling M (1986) Persistent inhibition of protein synthesis precedes delayed neuronal death in post ischemic gerbil hippocampus. Acta Neuropathol 71: 88–91
17. Yamasaki Y, Suzuki T, Yamaya H, Matsuura N, Onodera H, Kogure K (1992) Possible involvement of interleukin-1 in ischemic brain edema formation. Neurosci Lett 142: 45–47

Astroglial and Microglial Activation in Hippocampus of Rat After Global Forebrain Ischemia

K.-A. Hossmann and C. Wiessner

Summary

The astroglial and microglial reaction to ischemic neuronal death was studied in hippocampus of rats submitted to 30-min, near-complete forebrain ischemia. The global metabolic response to ischemia was assessed by measuring energy metabolism and protein synthesis, and the cell-specific reactions by using a battery of histological, in situ hybridization and immunocytochemical techniques.

Ischemia caused instantaneous suppression of electrophysiological function and global breakdown of energy metabolism, followed by rapid restoration of energy state and gradual recovery of electroencephalogram (EEG) upon recirculation. Protein synthesis was severely inhibited for several hours in all brain regions and remained permanently suppressed in the pyramidal neurons of the vulnerable CA1 sector. In this region neuronal death became histologically visible after 2–3 days.

Thirty min after the beginning of recirculation, the mRNAs of immediate-early genes (c-*jun*, c-*fos*) were expressed in neurons from all parts of the hippocampus, followed after about 1 h by the expression of various stress proteins such as heat shock protein 70 and glucose-regulated protein 78 (hsp70, grp78, respectively). After 3–7 days these changes normalized first in the resistant and later also in the vulnerable parts of the hippocampus. Expression of mRNAs of glia-associated proteins followed distinctly later. The message of astroglia-specific proteins such as glial fibrillary acidic protein and sulphated glycoprotein-2 (gfap, sgp-2, respectively) became visible between 6 h and 12 h, and that of microglia such as CR3 complement receptor β subunit and transforming growth factor β1 (CR3β, TGF-β1) from day 1 on. Glial protein products GFAP, vimentin, MUC 101, MUC 102, and OX-42, reached their maximum between 1 and 3 days and were still present after 7 days. Expression was most pronounced in CA1 sector but to a lesser degree also affected the resistant parts of hippocampus. These data suggest that gene products of reactive glia are not causally involved in neuronal death but are important components of the glial phenotype in regions with neuronal damage.

Max-Planck-Institute for Neurological Research, Department of Experimental Neurology, Gleueler Str. 50, 50931 Cologne, Germany

Maturation Phenomenon in Cerebral Ischemia II
U. Ito et al. (Eds.)

Introduction

Brief periods of global cerebral ischemia produce a delayed type of neuronal injury in selectively vulnerable areas, in particular, CA1 sector of hippocampus or the dorsolateral part of striate nucleus [13, 26]. In these regions neurons are irreversibly damaged, whereas glia and vascular cells survive. This status has been defined as "selective neuronal necrosis" [30]. Although the leading pathology of selective ischemic vulnerability is neuronal cell death, glial cells also undergo marked alterations [25, 29, 33]. In view of the multiple functions of glia, these alterations may be of pathophysiological significance for the manifestation of neuronal cell death. In particular, activated microglia could be capable of aggravating ischemic injury by the release of potentially cytotoxic substances, as suggested by the improvement of functional outcome with microglia-suppressing drugs [7]. Conversely, microglial proliferation may also have protective effects by shielding endangered neurons from their synaptic input, as shown after facial nerve transection [3]. It is also widely held that activated astroglia may support survival of injured neurons because it is a potent source of neurotrophic factors [23, 34].

In the following, we review data from our laboratory in which the astroglial and the microglial response to global cerebral ischemia were studied by immunocytochemical and in situ hybridization techniques to establish the relationship between global metabolic alterations and the evolution of neuronal cell damage in the selectively vulnerable hippocampus [14, 21, 36, 40].

Materials and Methods

Experiments were carried out in normothermic male Wistar rats, weighing between 220 g and 350 g. Global forebrain ischemia of 30-min duration was produced under halothane–nitrous oxide anesthesia by a modification of the four-vessel occlusion model of Pulsinelli and Brierly [27], as previously described by Schmidt-Kastner et al. [28]. On the first day both vertebral arteries were coagulated. On the following day, the carotid arteries were additionally occluded for 30 min. During ischemia electroencephalogram (EEG) was recorded to differentiate between animals with complete and incomplete ischemia. In some animals laser Doppler flow was also monitored to ascertain the relationship between electrophysiological and hemodynamic changes.

After ischemia, recirculation was restored by removing vessel clips and ascertaining free blood circulation through the carotid arteries by microscopic examination. Animals were allowed to recover for various periods ranging from 15 min to 7 days, reanesthetized and then either decapitated for freezing the brain in precooled isopentan, or perfusion fixed with buffered formaldehyde.

Histology

Neuronal damage in hippocampal formation was evaluated on cryostat or paraffin-embedded brain sections, using conventional cresyl-violet and hematoxylin-eosin stains.

In Situ Hybridization

Oligonucleotide probes specific for the mRNAs of immediate-early genes c-*fos* and c-*jun*, the stress proteins heat stress protein 70 (hsp70) and glucose-regulated protein 78 (grp78), the astroglia-associated proteins glial fibrillary acidic protein (GFAP) and sulphated glycoprotein (sgp)-2, as well as for the microglia-associated protein cr3β were synthesized on a Pharmacia Gene Assemble Special [21, 39].

Transforming growth factor (TGF)-β1 mRNA was investigated using a Riboprobe [39]. In situ hybridization was performed as described by Wisden et al. [41] with the following modifications: Unspecific background of brain sections was reduced by acetylation with acetic anhydride in triethanolamine and prehybridization for 30 min in prehybridization buffer [2 × standard sodium citrate (SSC), 50% formamide, 250 μg/ml sonified herring sperm DNA, 12.5 μg/ml poly(A)] at 40°C. Following dehydration 10 μl hybridization buffer [2 × SSC, 50% formamide, 10% dextrane sulfate, 100 μg/ml poly(A), 120 μg/ml heparin, 1 mg/ml herring sperm, DNA 5 m*M* DTT, 1 mg/ml bovine serum albumin] containing 1–5 pg/μl ^{35}S-labeled probe were applied to each brain slice and covered with a coverslip.

After 24-h hybridization at 40°C, hybridized radioactivity was visualized by film autoradiography (Hyperfilm β-max, Amersham, UK) with exposure times of 1–3 weeks. For cellular localization the glass slides were dipped in photoemulsion (LM-1, Amersham) and exposed for 1–3 weeks. The brain sections were lightly counterstained with cresyl-violet.

Immunocytochemistry

Activated astroglia was visualized by staining GFAP or vimentin using polyclonal rabbit anti-GFAP (DAKO, Hamburg, Germany) or monoclonal mouse anti-vimentin antibodies (Boehringer, Mannheim, Germany). Microglial activation was studied using the mouse monoclonal antibody OX-42, recognizing rat CR3 complement receptor, and the mouse monoclonal antibodies MUC 101 and 102, recognizing several microglia-specific proteins. Antibody detection was carried out using the peroxidase–antiperoxidase technique [29].

Nuclear Magnetic Resonance Spectroscopy

The effect of forebrain ischemia on energy metabolites was studied by ^{31}P nuclear magnetic resonance (NMR) spectroscopy. For this purpose, a remotely controlled carotid artery occlusion procedure was developed to allow induction and reversal of ischemia under continuous NMR recording. ^{31}P spectra were acquired with a time resolution of 1–5 min at 4.7 Tesla (Bruker, Karlsruhe, Germany) using a surface coil placed on the exposed skull [14]. Relative concentration changes of ATP and phosphocreatin (PCr) were calculated by peak integration and changes of pH by measuring the chemical shift between the PCr and inorganic phosphate peaks.

Protein Synthesis

Incorporation of [^{14}C]leucine in brain proteins was measured as described in detail by Widmann et al. [36]. In short, tracer was infused over 45 min, using a programmed infusion protocol to keep serum [^{14}C]leucine-specific activity constant throughout the infusion time. At the end of infusion, animals were decapitated and cryostat sections of brain were prepared at the level of dorsal hippocampus. Trichloroacetic acid (TCA)-soluble precursor radioactivity was removed by wash incubation, and protein-incorporated radioactivity was measured by quantitative autoradiography. The protein synthesis rate was measured from brain protein radioactivity and the integrated serum precursor-specific radioactivity, using the operational equation of Smith et al. [32] and appropriate precursor pool correction factors [36].

Results

General Pathophysiological Observations

Our modified four-vessel-occlusion model of rat produces near-complete forebrain ischemia for 30 min, followed by rapid restoration of blood flow after removal of vessel clips (Fig. 1). EEG was suppressed within less than 30 s after onset of ischemia

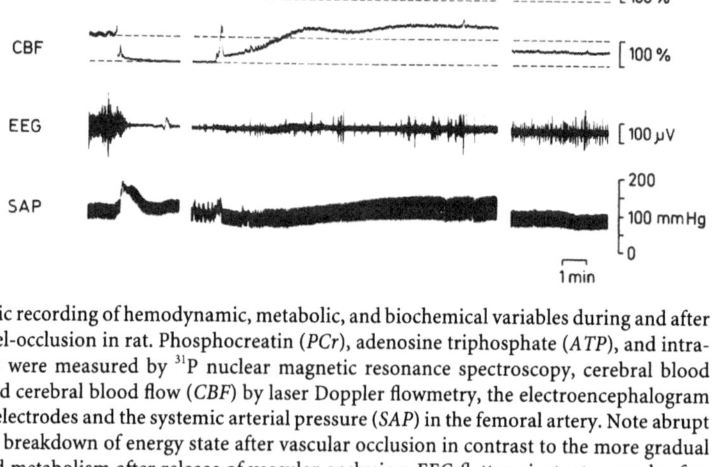

Fig. 1. Polygraphic recording of hemodynamic, metabolic, and biochemical variables during and after 30-min four-vessel-occlusion in rat. Phosphocreatin (*PCr*), adenosine triphosphate (*ATP*), and intracellular pH (*pHi*) were measured by ^{31}P nuclear magnetic resonance spectroscopy, cerebral blood volume (*CBV*) and cerebral blood flow (*CBF*) by laser Doppler flowmetry, the electroencephalogram (*EEG*) with scalp electrodes and the systemic arterial pressure (*SAP*) in the femoral artery. Note abrupt arrest of CBF and breakdown of energy state after vascular occlusion in contrast to the more gradual return of flow and metabolism after release of vascular occlusion. EEG flattens instantaneously after onset of ischemia, and it gradually recovers in parallel to the restoration of flow and metabolism (reproduced from [14])

and began to recover within 30 min after the beginning of recirculation. Animals with residual blood flow after carotid artery occlusion showed delayed suppression of EEG activity and were excluded from further analysis.

NMR-spectroscopical measurements of energy metabolites and intracellular pH confirmed the almost immediate breakdown of energy-producing metabolism, and the gradual recovery in parallel with restoration of blood flow during recirculation (Fig. 1). Following 30-min forebrain ischemia, animals remained in an unresponsive state for 1–3 h. Then animals got up and maintained a characteristic posture for the next 24 h. The lumbar spine was flexed, and the animals periodically searched the ground by chewing, biting, and moving the forepaws. These stereotypes could be interrupted by acoustic stimuli, while visual responses were sluggish. Otherwise, there were no consistent neurological symptoms. By the second day, posture and overall behavior tended to normalize, and most animals were able to walk, feed, and drink. During the initial 24-h recirculation, epileptic seizures were absent, but after longer survival times convulsions occasionally occurred.

Histological Evaluation of Neuronal Damage

The pattern and time course of neuronal death following 30-min forebrain ischemia was very consistent. Loss of neurons in the dorsolateral caudate putamen became evident within 12 h of recirculation, whereas cell death in the CA1 sector of the hippocampus occurred after 2–3 days recirculation. Some disseminated neuronal lesions were also observed in layers 3 and 5 or 6 of the cortex, and the ventral thalamus. In the following, our description will focus on the changes observed in hippocampus.

Protein Synthesis

In control animals, protein synthesis reflected the typical relationship to neuronal density with high synthesis rates in grey and low synthesis rates in white matter structures (Fig. 2).

During the initial 2 h of recirculation, [^{14}C]leucine incorporation into gray matter proteins declined from about 15 nmol/100 g/min to less than 10%, but subsequently it slowly recovered, except in the selectively vulnerable areas such as caudate putamen or the pyramidal layer of the CA1 sector of hippocampus where it remained severely suppressed (Fig. 2). After 7 days high protein synthesis rate was observed in activated astroglia and microglia of the CA1 sector and other regions with neuronal injury, resulting in the apparent return of the synthesis rate to near control levels.

In Situ Hybridization Studies

The genomic response of hippocampus following 30-min global ischemia varied greatly, depending on the kind and topographical distribution of the investigated mRNAs (Tables 1–3).

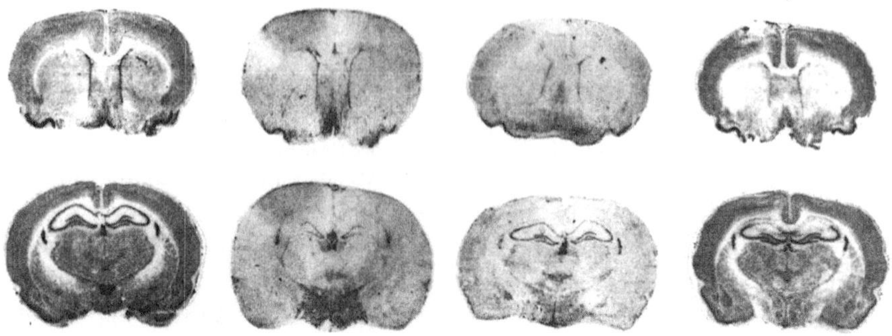

Fig. 2. Autoradiographic imaging of [^{14}C]leucine incorporation into brain proteins before (control), and at various times (1 h–3 days) after 30-min four-vessel-occlusion in rat. Two coronal sections are shown passing through the caudate putamen and dorsal hippocampus, respectively. Note global suppression of protein synthesis during the initial 6 h of recirculation, followed by recovery in most brain regions except caudate putamen and CA1 sector of hippocampus (modified from [36])

The mRNAs of immediate-early genes (IEGs) c-*fos* and c-*jun* were activated throughout the hippocampus as early as 30 min after the onset of recirculation, and reached a maximum after about 3 h. Subsequently, mRNA induction gradually declined and disappeared at between 3 and 7 days. Normalization of IEG induction was slightly slower in the vulnerable CA1 sector than in the resistant CA3 sector resulting, in some animals, in the apparently selective expression in the CA1 sector after 2 days (Fig. 3). Expression of stress protein mRNAs hsp70 and grp78 was similar to that of IEGs, but became prominent only from 1 h on. Stress proteins were also expressed in all hippocampal sectors with a gradual return to normal between 3 and 7 days. Here again, return to normal was delayed in CA1 sector, as shown after 2 days of recirculation (Fig. 3). The constitutively expressed grp78 mRNA was markedly suppressed in the CA1 sector from day 3 on in parallel with the morphological disintegration of pyramidal neurons in this sector (Fig. 3).

Induction of astroglia-associated mRNAs, gfap and sgp-2, followed distinctly later than that of IEGs or stress proteins. An increase in mRNA expression became visible between 6 and 12 h and was prominent after 1–2 days. In the CA1 sector the maximum expression was reached on day 7, whereas in the resistant CA3 sector expression already began to decline at that time.

Microglia-associated mRNAs coding for the CR3 complement receptor β subunit (CR3β) and TGF-β1 proteins were expressed last in the sequel of genomic responses. Changes became visible only from day 3 on and were strictly confined to regions of established cell injury, i.e., in CA1 sector and other parts of the brain with histologically detectable neuronal death.

Immunohistochemical Studies

Astroglial activation was investigated by visualizing the changes in GFAP and vimentin concentrations. In the vulnerable CA1 sector increased GFAP immunoreac-

Table 1. Marker genes for neuronal and glial activation

Gene	Class	Function of gene product	Post-ischemic induction
c-fos	Immediate-early gene	Transcription factor (forms heterodimers with c-jun)	Mostly neuronal
c-jun	Immediate-early gene	transcription factor	Basal expression in various cell types of the central nervous system, induction mostly neuronal
hsp70	Heat shock protein	Protein folding, refolding of denatured proteins	Mostly neuronal
grp78	Glucose-regulated protein	Protein folding in ER, refolding of denatured protein in the ER	Ubiquituous basal expression, induction mostly neuronal
gfap (glial fibrillary acidic protein)	Intermediate filament	Component of the astrocytic cytoskeleton	Basal expression in certain populations of astrocytes, induction specific for astrocytes
sgp-2 (sulfated glycoprotein-2, clusterin)	Secreted glycoprotein	Not precisely defined, membrane stability, complement regulation	Ubiquituous basal expression, induction strictly in astrocytes
cr3β (CR3 complement receptor β subunit)	Integrin	Cell adhesion in complement-mediated cell lysis	Expressed in cells of monocytic origin, induction in microglia
TGF-β1 (transforming growth fator-β1)	Cytokine	Manifold functions in cell growth and division, antagonizes tumor necrosis factor-α effects, neuroprotective	Induction mostly in microglia

ER, endoplasmic reticulum.

Table 2. Neuronal and glial response in vulnerable CA1 sector of hippocampus after 30-min forebrain ischemia of rat

	Recirculation times									
	15 min	30 min	1 h	3 h	6 h	12 h	1 day	2 days	3 days	7 days
Neuronal response										
Protein synthesis inhibition	+++	+++	+++	+++	+++	+++	+++	+++	+++	+++
c-fos, c-jun mRNAs	–	++	++	+++	++	++	++	++	(+)	–
hsp70, grp78 mRNAs	–	+	++	+++	+++	+++	+++	++	–	–
Ischemic cell change	–	–	–	–	–	–	(+)	+	++	+++
Astroglial response										
gfap, sgp-2 mRNAs	–	–	–	–	(+)	+	++	++	+++	+++
GFAP	–	–	–	–	–	–	+	++	+++	+++
Vimentin	–	–	–	–	–	–	–	+	++	+++
Microglial response										
cr3β, TGF-β1 mRNAs	–	–	–	–	–	–	–	–	++	++
OX-42, MUC 101, MUC 102	–	–	–	–	–	–	+	+	++	+++

Intensity changes are evaluated qualitatively: – no change; + slight change; ++ distinct change; +++ massive change.

Table 3. Neuronal and glial response in resistant CA3 sector of hippocampus after 30-min forebrain ischemia of rat

	Recirculation time									
	15 min	30 min	1 h	3 h	6 h	12 h	1 day	2 days	3 days	7 days
Neuronal response										
Protein synthesis inhibition	+++	+++	+++	++	+	–	–	–	–	–
c-fos, c-jun mRNAs	–	++	++	+++	++	++	++	+	+	(+)
hsp70, grp78 mRNAs	–	+	++	++	+++	+++	+++	–	–	–
Ischemic cell change	–	–	–	–	–	–	–	–	–	–
Astroglial response										
gfap, sgp-2 mRNAs	–	–	–	–	(+)	+	++	++	++	+
GFAP	–	–	–	–	–	–	–	–	+	+
Vimentin	–	–	–	–	–	–	–	–	–	–
Microglial response										
cr3β, TGF-β1 mRNAs	–	–	–	–	–	–	–	–	–	–
OX-42, MUC 101, MUC 102	–	–	–	–	–	–	+	+	+	–

Intensity changes are evaluated qualitatively: –, no change; +, slight change; ++, distinct change; +++, massive change.

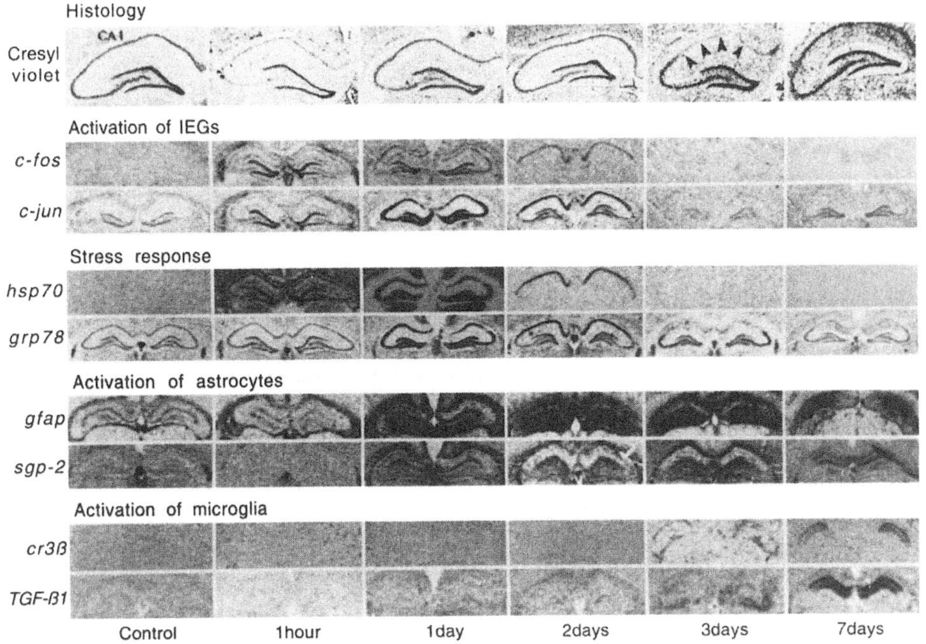

Fig. 3. Comparison of histological alterations in hippocampus with the regional expression pattern of various mRNAs before (control) and at different times (1 h–7 days) after 30-min four-vessel-occlusion in rat. The histological sections were stained with cresyl-violet, and the distribution of mRNAs was assessed by in situ hybridization. Note early expression of immediate-early genes and stress proteins in neurons as compared to the more delayed expression of the mRNAs of astroglia- and microglia-associated genes. Neuronal loss in CA1 sector of hippocampus is visible after 3 days (*arrows*). The apparent return of staining in this area after 7 days is due to the massive proliferation of microglia. *TGF-β1*, transforming growth factor-β1

tivity was visible after 1 day and became very strong from 3 days onwards (Fig. 4). Vimentin immunoreactivity appeared after 2 days and became prominent between 3 and 7 days. In the resistant CA3 sector astroglial response was much less pronounced. There was some faint immunoreactivity of GFAP between 3 and 7 days but expression of vimentin was strictly confined to the regions with established neuronal injury (Tables 2, 3).

Microglial activation, as studied by the immunoreactivity of the microglia-specific antigens OX-42, MUC 101 and MUC 102, started even later than the astroglial response (Tables 2, 3). In the vulnerable CA1 sector faint staining became visible after 1 day and then it gradually increased until immunoreactivity reached a maximum after 7 days (Fig. 5). In the resistant CA3 sector increase in immunoreactivity of microglia-specific proteins was modest and was detected only between the first and third day of recirculation.

Discussion

The comparison of the temporal and topical evolution of the neuronal and glial alterations in hippocampus after 30-min, near-complete forebrain ischemia provides

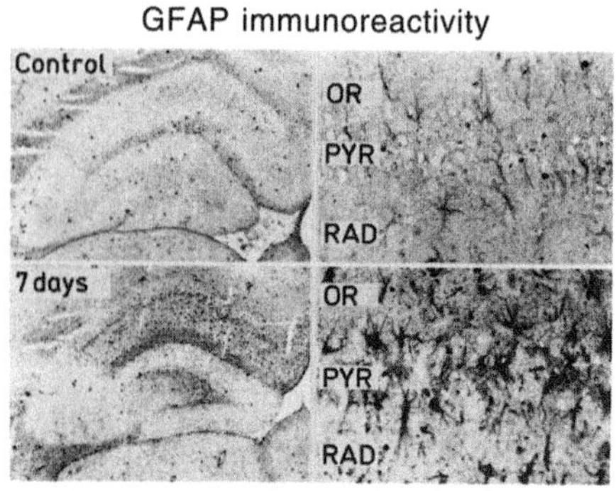

Fig. 4. Immunohistochemical staining of astroglia with GFAP antibodies in hippocampus of rat before (control) and at 7 days after 30-min four-vessel-occlusion of rat. Higher magnification (*right*) shows that the increase in GFAP immunoreactivity in CA1 sector is related to the hypertrophy of reactive astrocytes. *OR*, stratum oriens; *PYR*, stratum pyramidale; *RAD*, stratum radiatum

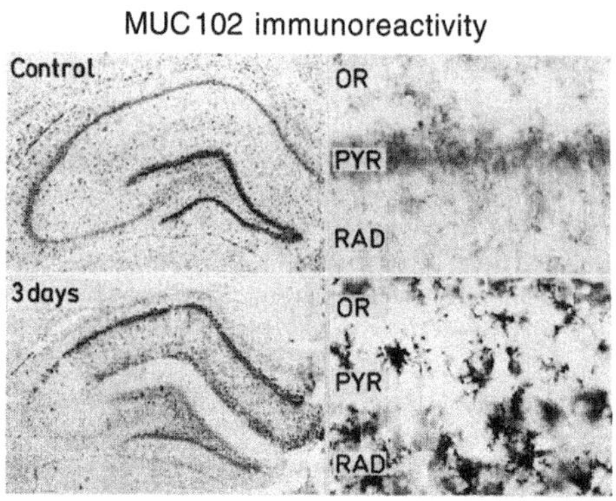

Fig. 5. Immunohistochemical staining of microglia with MUC 102 antibody before (control) and at 3 days after 30-min four-vessel occlusion of rat. Sections are slightly counterstained with cresyl-violet. Note massive staining of reactive microglia in CA1 sector at 3 days after ischemia. Higher magnification (*right*) shows vertical rod cell arrangement. *OR*, stratum oriens; *PYR*, stratum pyramidale; *RAD*, stratum radiatum

a rather clear picture of the dynamics of ischemic injury. The early neuronal reaction to ischemia is undoubtedly the dominating change. The mRNAs of immediate-early genes are induced in pyramidal neurons as early as 30 min after the restoration of blood flow, followed within 1 h by a prominent induction of the mRNAs of stress proteins. The astroglial reaction, as reflected by the induction of the mRNAs of GFAP and sgp-2, is delayed to about 12 h, and the induction of CR3β and TGF-β1 mRNAs as markers of the microglial reaction follows even later. This sequel conforms essentially to similar observations made in other laboratories, although the timing of the glial induction may vary, depending on the animal species, the duration of ischemia, and the selection of marker proteins used [1, 10, 12, 16–18, 24, 31]. In particular, earlier activations have been shown using isolectin as the microglial marker [10, 19].

An important biochemical change that modulates the translation of the genomic message is the general postischemic suppression of the protein synthesizing machinery. This suppression lasts much longer than that of the energy metabolism, and it is irreversible in neurons destined to die after the ischemic impact [8]. The reason for this suppression has not been fully established but there are strong indications that it is due to the selective inhibition of polypeptide chain initiation in response to the inactivation of one or several of the initiation factors [4, 9]. The global inhibition of protein synthesis is the reason for the fact that, despite upregulation of the message, certain proteins are not or only tardily formed. This is particularly true for the protein products of IEGs and stress proteins, the synthesis of which is greatly delayed [21, 35] or even permanently suppressed, as in the CA1 sector of gerbil [22]. As far as the glial response is concerned, postischemic inhibition of global protein synthesis seems to be of less concern. The genomic message for the proteins investigated here is not expressed before 1 day after ischemia, i.e., at a time at which global protein synthesis has largely recovered [36]. The late activation of glial cells is, therefore, due to processes which occur late in the pathophysiological cascade of postischemic reperfusion and which must be related to secondary events that are indirectly linked to the initial ischemic impact.

The most obvious reason is the delayed disintegration of neurons in selectively vulnerable brain areas which, in fact, correlates closely with the activation of microglia [10, 18]. This activation is associated with the production of nitric oxide [5], leading first to the upregulation of major histocompatibility complex (MHC) antigens and later to phagocytosis of the disintegrated neurons [6]. In contrast to astrocytes, microglia undergo pronounced cell proliferation in response to neuronal degeneration which is reflected by a transient upregulation of the mRNA for the cell-cycle upregulating protein cyclin D1 [38].

As far as the astroglial activation is concerned, a correlation with the manifestation of neuronal death is less evident. As shown in Tables 2 and 3, the induction of astroglia-associated proteins begins well before the manifestation of neuronal cell death, and it transiently expands into the resistant CA3 sector in which neuronal injury is absent. This raises the question of whether the astroglia receives signals from the ischemia-stressed neurons before cells begin to disintegrate or if it responds to pathological changes that are indirectly linked to the neuronal injury.

In a previous study from our laboratory, Schmidt-Kastner et al. [29] compared activation of GFAP with blood–brain barrier injury and found a rather close correlation. In fact, extravasation of serum proteins could be detected in the vulnerable regions before neuronal injury became visible, and it spread to some degree to the noninjured resistant parts of the hippocampus. It has been shown that vasogenic edema is associated with an activation of polyamine metabolism [15], and that the polyamine putrescine exerts glia-stimulating activity [20]. It is, therefore, conceivable that the induction of GFAP is mediated by edema formation rather than by neuronal injury. Interestingly, the expression of vimentin occurred later than that of GFAP and was strictly confined to regions with manifested neuronal injury. The same is true for an increase of the amyloid precursor protein (APP) immunoreactivity which is localized mostly in reactive astrocytes of the lesioned hippocampus [2, 24]. This suggests that the astroglia responds to the different elements of the pathophysiological cascade in a very specific way.

In conclusion, the present study of astro- and microglial activation in rat hippocampus after 30-min forebrain ischemia clearly demonstrates that the glial reaction is most pronounced in areas with established neuronal injury, either in response to the formation of vasogenic edema or to the release of mediators from the injured neurons. Our observations do not lend support to the notion that microglial activation is a causal factor of delayed neuronal death, but they stress the importance of a well-organized differentiated glial response as an integral part of the pathological process.

References

1. Araki T, Kato H, Liu XH, Kogure K, Itoyama Y (1994) Induction of heat shock protein 70 and glial fibrillary acidic protein in the postischemic gerbil hippocampus. Metab Brain Dis 9: 369–375
2. Banati RB, Gehrmann J, Wießner C, Hossmann K-A, Kreutzberg GW (1995) Glial expression of the ß-amyloid precursor protein (APP) in global ischemia. J Cereb Blood Flow Metab 15: 647–654
3. Blinzinger K, Kreutzberg GW (1968) Displacement of synaptic terminals from regenerating motoneurons by microglial cells. Z Zellforsch 85: 145–157
4. Cooper HK, Zalewska T, Kawakami S, Hossmann K-A, Kleihues P (1977) The effect of ischaemia and recirculation on protein synthesis in the rat brain. J Neurochem 28: 929–934
5. Faraci FM, Brian JE (1994) Nitric oxide and the cerebral circulation. Stroke 25: 692–703
6. Finsen BR, Jorgensen MB, Diemer NH, Zimmer J (1993) Microglial MHC antigen expression after ischemic and kainic acid lesions of the adult rat hippocampus. Glia 7: 41–49
7. Giulian D, Vaca K (1993) Inflammatory glia mediate delayed neuronal damage after ischemia in the central nervous system. Stroke 24: I84–I90
8. Hossmann K-A (1993) Disturbances of cerebral protein synthesis and ischemic cell death. Prog Brain Res 96: 161–177
9. Hu BR, Wieloch T (1993) Stress-induced inhibition of protein-synthesis initiation – modulation of initiation factor-ii and guanine-nucleotide exchange factor activities following transient cerebral-ischemia in the rat. J Neurosci 13: 1830–1838
10. Ivacko JA, Sun R, Silverstein FS (1996) Hypoxic-ischemic brain injury induces an acute microglial reaction in perinatal rats. Pediatr Res 39: 39–47
11. Kato H, Kogure K, Araki T, Itoyama Y (1995) Graded expression of immunomolecules on activated microglia in the hippocampus following ischemia in a rat model of ischemic tolerance. J Cereb Blood Flow Metab 11: 966–973
12. Kindy MS, Bhat AN, Bhat NR (1992) Transient ischemia stimulates glial fibrillary acid protein and vimentin gene expression in the gerbil neocortex, striatum and hippocampus. Mol Brain Res 13: 199–206
13. Kirino T, Tamura A, Sano K (1985) Selective vulnerability of the hippocampus to ischemia. Reversible and irreversible types of ischemic cell damage. Prog Brain Res 63: 39–58
14. Kloiber O, Miyazawa T, Hoehn-Berlage M, Hossmann K-A (1993) Simultaneous ^{31}P NMR spectroscopy and laser Doppler flowmetry of rat-brain during global-ischemia and reperfusion. NMR Biomed 6: 144–152
15. Koenig H, Goldstone AD, Lu CY (1989a) Blood-brain barrier breakdown in cold-injured brain is linked to a biphasic stimulation of ornithine decarboxalase activity and polyamine synthesis: both are coordinately inhibited by verapamil, dexamethasone, and aspirin. J Neurochem 52: 101–109
16. Kogure K, Kato H (1993) Altered gene expression in cerebral ischemia. Stroke 24: 2121–2127

17. Lehrmann E, Kiefer R, Finsen B, Diemer NH, Zimmer J, Hartung HP (1995) Cytokines in cerebral ischemia: expression of transforming growth factor beta-1 (TGF-beta 1) mRNA in the postischemic adult rat hippocampus. Exp Neurol 131: 114–123
18. McRae A, Gilland E, Bona E, Hagberg H (1995) Microglia activation after neonatal hypoxic-ischemia. Dev Brain Res 84: 245–252
19. Morioka T, Kalehua AN, Streit WJ (1991) The microglial reaction in the rat dorsal hippocampus following transient forebrain ischemia. J Cereb Blood Flow Metab 11: 966–973
20. Morrison RS, De Vellis J, Lee YL, Bradshaw RA, Eng LF (1985) Hormones and growth factors induce the synthesis of glial fibrillary acidic protein in rat brain astrocytes. J Neurosci Res 14: 167–176
21. Neumann-Haefelin T, Wiessner C, Vogel P, Back T, Hossmann K-A (1994) Differential expression of the immediate early genes c-fos, c-jun, junB, and NGFI-B in the rat brain following transient forebrain ischemia. J Cereb Blood Flow Metab 14: 206–216
22. Nowak TS (1990) Protein synthesis and the heat shock/stress response after ischemia. Cereb Brain Metab Rev 2: 345–366
23. Oderfeld-Nowak B, Bacia A, Gradkowska M, Fusco M, Vantini G, Leon A, Aloe L (1992) In vivo activated brain astrocytes may produce and secrete nerve growth factor-like molecules. Neurochem Int 21: 455–461
24. Palacios G, Mengod G, Tortosa A, Ferrer I, Palacios JM (1995) Increased beta-amyloid precursor protein expression in astrocytes in the gerbil hippocampus following ischaemia: association with proliferation of astrocytes. Eur J Neurosci 7: 501–510
25. Petito CK, Halaby IA (1993) Relationship between ischemia and ischemic neuronal necrosis to astrocyte expression of glial fibrillary acidic protein. Int J Dev Neurosci 11: 239–247
26. Pulsinelli WA, Brierley JB (1979) A new model of bilateral hemispheric ischemia in the unanesthetized rat. Stroke 10: 267–272
27. Pulsinelli WA, Brierley JB, Plum F (1982) Temporal profile of neuronal damage in a model of transient forebrain ischemia. Ann Neurol 11: 491–498
28. Schmidt-Kastner R, Paschen W, Große Ophoff B, Hossmann K-A (1989) A modified 4-vessel occlusion model for inducing incomplete forebrain ischemia in rats. Stroke 20: 938–946
29. Schmidt-Kastner R, Szymas J, Hossmann K-A (1990) Immunohistochemical study of glial reaction and serum-protein extravasation in relation to neuronal damage in rat hippocampus after ischemia. Neuroscience 38: 527–540
30. Scholz W (1953) Selective neuronal necrosis and its topistic patterns in hypoxemia and oligemia. J Neuropathol 12: 249–261
31. Sharp FR, Sagar SM (1994) Alterations in gene expression as an index of neuronal injury: heat shock and the immediate early gene response. Neurotoxicology 15: 51–59
32. Smith CB, Deibler GE, Eng K, Schmidt K, Sokoloff L (1988) Measurement of local cerebral protein synthesis in vivo: influence of recycling of amino acids derived from protein degradation. Proc Natl Acad Sci USA 85: 9341–9345
33. Streit WJ (1993) Microglial-neuronal interactions. J Chem Neuroanat 6: 261–266
34. Takami K, Kiyota Y, Iwane M, Miyamoto M, Tsukuda R, Igarasi K, Shino A, Wanaka A, Shiosaka S, Tohyama M (1993) Upregulation of fibroblast growth factor-receptor messenger RNA expression in rat brain following transient forebrain ischemia. Exp Brain Res 97: 185–194
35. Tomioka C, Nishioka K, Kogure K (1993) A comparison of induced heat-shock protein in neurons destined to survive and those destined to die after transient ischemia in rats. Brain Res 612: 216–220
36. Widmann R, Miyazawa T, Hossmann K-A (1993) Protective effect of hypothermia on hippocampal injury after 30 minutes of forebrain ischemia in rats is mediated by postischemic recovery of protein-synthesis. J Neurochem 61: 200–209
37. Wiessner C, Back T, Bonnekoh P, Kohno K, Gehrmann J, Hossmann K-A (1993) Sulfated glycoprotein-2 mRNA in the rat brain following transient forebrain ischemia. Mol Brain Res 20: 345–352
38. Wiessner C, Brink I, Lorenz P, Neumann-Haefelin T, Vogel P, Yamashita K (1996) Cyclin D1 messenger RNA is induced in microglia rather than in neurons following transient forebrain ischaemia. Neuroscience (in press)
39. Wiessner C, Gehrmann J, Lindholm D, Topper R, Kreutzberg GW, Hossmann K-A (1993) Expression of transforming growth factor-β1 and interleukin-1β mRNA in rat brain following transient forebrain ischemia. Acta Neuropathol (Berl) 86: 439–446

40. Wiessner C, Vogel P, Neumann-Haefelin T, Hossmann K-A (1995) Molecular correlates of delayed neuronal death following transient forebrain ischemia in the rat. Acta Neurochir (Wien) 66 [Suppl]: 1–7
41. Wisden W, Morris BJ, Hunt SP (1991) In situ hybridization with synthetic DNA probes. In: Chad J, Wheal H (eds) Molecular neurobiology – a practical approach. Oxford University Press, Oxford, pp 205–228

Glial Protection Against Neuronal Damage

O. Kempski and C. Volk

Summary

Glial homeostatic mechanisms are involved in neuronal protection during the early phase of cerebral ischemia. These protective effects include, among others, glutamate uptake and the regulation of pH in the extracellular space of the brain. Uptake of glutamate goes along with glial swelling, as does the elimination of protons from the glial cytosol. Five transport systems interact in order to maintain a normal intra- and extracellular pH in the brain.

From our in vitro data we conclude that in vivo these transport systems are heterogeneously distributed over the glial membrane, thereby allowing for a compartmental buffering function of glia. During reperfusion the action of these homeostatic mechanisms may aggravate cytotoxic glial swelling, and thereby impede reperfusion.

Introduction

During an ischemic episode a multitude of extra- and intracellular parameters are rapidly changing since pH, K_e^+, free fatty acids, or transmitters such as glutamate are present. These changes are believed to aggravate nerve and glial swelling and to eventually participate in nerve cell death. We have accumulated a host of data supporting a protective, homeostatic role of astroglia during the early phase of ischemia. So far the effects of anoxia, ouabain, arachidonic acid, glutamate, elevated K^+ and acidosis have been studied. In this short review results from those studies will be briefly summarized, with special focus on the regulation of intracellular pH.

Materials and Methods

All experiments were conducted in vitro using either astrocytes from primary culture or C6 glioma cells. Experiments are performed under close control of the extracellular environment enabling one to selectively modify given parameters [8–13, 20, 24]. Cell volume was measured in suspended cells using flow cytometry. Intracellular pH (pH_i) was determined indirectly by bis-carboxyethyl carboxyfluorescein (BCECF)

Institute for Neurosurgical Pathophysiology, Johannes Gutenberg-University Mainz, 55101 Mainz, Germany

Maturation Phenomenon in Cerebral Ischemia II
U. Ito et al. (Eds.)

fluorescence emitted after stimulation at 450 nm and 495 nm. Intracellular acidosis was induced by an acid load with 25 m*M* NH_4Cl, and extracellular acidosis ($pH_e = 6.4$) is studied in either hydroxyethylpiperazine ethanesulfonic acid (HEPES) or bicarbonate buffered media.

Results

Astrocytes as well as C6 cells possess at least five transport systems involved in pH control: Under physiological conditions H^+ adenosine triphosphatase (ATPases) actively extrude protons; Na^+-HCO_3^--cotransport, Na^+-dependent as well as Na^+-independent Cl^-/HCO_3^--antiport participate, depending on membrane potential and the presence of bicarbonate. With decreasing pH_i the Na^+/H^+-antiporter becomes activated once a threshold of $pH_i = 6.8–6.6$ is reached. These conclusions are drawn from inhibition experiments using either *N*-ethylmaleinimide (NEM) (1 m*M*) or nitrobenzoxadiazole (NBD)-Cl (100 μM) to block H^+-ATPases, amiloride derivatives to inhibit Na^+/H^+-antiport, SITS or DIDS to block Na^+-HCO_3^--cotransport and Cl^-/HCO_3^--antiport, or Na^+- or HCO_3^--free media. At extracellular acidosis these mechanisms in vitro are not sufficient to maintain a normal pH_i, however. Although pH-regulatory systems are activated and cell volume increases, the pH_i within 30 min nearly equilibrates with pH_e at 6.4. The presence of H^+-ATPases in astrocytes as well as C6 cells is concluded from experiments with NEM or NBD-Cl where pH_i gradually dropped to 6.6–6.7 and then reached a stable plateau, indicating activation of Na^+/H^+-exchange at this threshold. Stimulation of Na^+/H^+-exchange with phorbol esters (50 n*M*) leads to an elevation of this activation threshold. In Na^+-free media acidification is far more pronounced. The reduction of cellular energy reserves by iodoacetate (2.5 m*M*) also goes along with gradual acidification to a plateau at $pH_i = 6.7$ as an indication of failing H^+-ATPases but still functioning Na^+/H^+-exchange.

Discussion

Glial pH-Regulation

So far the existence of at least four pH-regulatory systems in astrocytes was well accepted [22]: The Na^+/H^+-antiporter and Na^+-dependent Cl^-/HCO_3^--antiport were considered most important for pH-regulation. Na^+-HCO_3^--cotransport causes alkalinization after glial depolarization, whereas Na^+-independent Cl^-/HCO_3^--antiport is thought to become active after intracellular alkalinization. Our observation that glial cells are equipped with H^+-ATPases confirms earlier data [18] and is important since the activation of an H^+-ATPase does not go along with the influx of osmotic activity as seen after activation of the Na^+/H^+-antiport [24]. In other words: only if energy reserves to support the function of H^+-ATPases are exhausted will pH_i drop below the activation threshold of the Na^+/H^+-antiporter. Now glial swelling occurs, with the remaining sodium gradient serving as a "battery" in a final attempt to prevent tissue acidosis.

Glial Swelling in Acidosis

Lactacidosis from anaerobic metabolism is regularly found during cerebral ischemia, seizures and head injury [23]. A marked decrease of brain tissue pH has been demonstrated in cerebral ischemia, particularly in hyperglycemic subjects, where lactic acid may accumulate to 20–30 m*M* [7, 19]. Tissue pH may drop down to pH 5.5 [4] or even lower. Acidosis has long been suspected as a mediator of brain damage [23]. Using our in vitro system we were able to confirm the swelling-inducing capacity of extracellular acidosis induced either by a strong inorganic acid [10] or by lactic acid [24]. A pH range between pH 7.6 and pH 4.2 was studied under strict maintenance of isotonicity. Glial volume was found to increase if the extracellular pH was titrated to 6.8 or below. From this level downward, the extent of swelling depended on the degree of acidosis and the duration of exposure. So, lactacidosis of pH 6.2 for 60 min led to a 24.5% increase in cell volume, while pH 5.0 or pH 4.2 increased cell size for 51.1% or 90.9%, respectively [24]. Cell viability was little affected down to pH 6.2. Cell swelling could be inhibited by replacement of Na^+ and bicarbonate in the medium by use of choline chloride and HEPES as buffer.

From these results acidosis-induced swelling was explained as a consequence of mechanisms to maintain a normal intracellular pH: Addition of acid to bicarbonate buffered media leads to the formation of carbonic acid which immediately dissociates to CO_2 and water. In turn, intracellular acidosis can develop, since CO_2 freely passes the cell membrane, and, catalyzed by carbonic anhydrase, again forms carbonic acid. To prevent acidosis, the resulting H^+-ions are exchanged against sodium ions. A resulting influx of Na^+ and Cl^- ions accompanied by water explains swelling.

This concept implies continuation of pH regulation even under conditions of severe extracellular acidosis. Opposed to this are observations of Kraig and Chesler [14], who found glial pH_i to decrease tremendously – and far below total tissue pH – during hyperglycemic ischemia in vivo. The authors explain their observation by a combination of three mechanisms: (1) the continued acid production in glia, (2) slow acid efflux as a result of maintained membrane integrity, and (3) acid inhibition of antiport mechanisms. Obviously the two proposed mechanisms do not fit together: The continued, amiloride-inhibitable swelling at pH levels as low as 4.6 speaks against an inhibition of the antiports by low pH. Second, a low extracellular pH is not necessarily "suicidal" for glia, which in vitro may survive pH 5.6 for more than 30 min.

Our current results prove that, in contrast to our original concept, severe extracellular acidosis is in fact rapidly followed by intracellular acidification, which occurs even faster in the presence of bicarbonate buffer. The latter at least fits the concept that CO_2 may enter the cell freely to form carbonic acid, and then to dissociate to protons and bicarbonate. The intriguing observation, however, is a gradual acidification for 30 min nearly down to the extracellular pH of 6.4 in HEPES buffer. Although Na^+/H^+ exchange is activated by intracellular acidification – demonstrated by acid loading experiments – this does not suffice to maintain a normal pH in extracellular acidosis.

The results enable us to expand the model of the glial cell as a control organ for the cerebral extracellular space. It is based on a proposed *functional polarization* of astrocytes in vivo, with ion channels and transport systems unevenly distributed over the cell membrane (or alternatively, activated unevenly) enabling a "compartmental buffering" of the extracellular pH. The hypothetic model is similar to mechanisms proposed by Newman [17] and Kraig et al. [15], and could explain many of the observed phenomena: Under physiological conditions H^+-ATPases would prevent the influx of acid equivalents to a Donnan equilibrium, i.e., to a calculated pH_i of 6.3 in astrocytes. With enhanced neuronal activity glial cells would depolarize and reduce the normally outward directed Na^+/HCO_3^--cotransport. This could explain the often observed glial alkalization with neuronal activity and would also cause a pH-induced increase of local cerebral blood flow (LCBF) to remove acid equivalents and to increase the supply of nutrients. In case of acid production by neurons, a transient acid entry into glia would permit a rapid "enlargement" of the virtual buffer space if the interior of the astrocyte is accessible for H^+-ions. These would predominantly be extruded towards the capillary bed by respective transporters as long as there is blood flow to remove acid equivalents. For the Na^+/HCO_3^--cotransporter such a unilateral localization has been proposed at least for the retina [17]. In cases where acid is not removed by H^+-ATPases but rather via Na^+/H^+-exchange or Na^+/HCO_3^--cotransport an elevation of intracellular sodium is the consequence. The resulting osmotic load could explain glial swelling as observed in vitro.

This model is in line with the neuropathological observation that cytotoxic glial swelling is first observed in glial endfeet around small blood vessels. So far it was not clear why cells which would receive nutrients and oxygen most easily should suffer first. With the proposed model glial endfeet should be rich in Na^+/H^+-exchange or Na^+/HCO_3^--cotransport systems: in conditions of ATP shortage but a remaining sodium gradient, glial endfeet would accumulate sodium, and hence osmoles to explain early swelling. Likewise the model would explain why amiloride analogues interfere with cytotoxic swelling.

Glutamate

Glutamate is an excitatory transmitter in the central nervous system (CNS) and, under pathophysiologic conditions, assumes neurotoxic properties. Glutamate is ubiquitous in the CNS but strictly compartmentalized in the intracellular space with intracellular concentrations up to 15 m*M*, more than any other free amino acid in the brain. The neurotoxic potential of glutamic acid has been proven for cerebral ischemia and brain injury and is commonly termed excitotoxicity. It is now generally accepted that glutamate may destroy neurons by excessive activation of excitatory receptors. On the other hand, glutamate is the most important excitatory transmitter in the CNS. Hence, under physiological conditions glutamate is highly compartmentalized, and there are efficient uptake systems both in the glial and in the neuronal membrane to clear the extracellular space from glutamate [2, 3, 5, 6, 21]. This process is coupled to a Na^+-downhill influx into the cell and thus is energy-dependent. "High-affinity" sodium-dependent glutamate uptake has been known about for more than 20 years, and recently transporter molecules have been cloned. Glial glutamate uptake is accompanied by an inward membrane current which is abolished if extracellular sodium or potassium are removed. Respective studies conclude that glutamate is taken up by an electrogenic uptake carrier which cotransports one molecule of glutamate together with two to three sodium ions into the cell and countertransports one potassium ion out of the cell [2, 3]. The ATP-dependent transformation of glutamate to glutamine via the glial enzyme glutamine synthetase may accelerate energy expenditure in energy-deprived conditions.

Even a still operative glial glutamte uptake in the vicinity of an ischemic focus, i.e., in the ischemic penumbra, may compromise central nervous function by causing glial swelling. Swelling is a direct consequence of glutamate accumulation together with Na^+-ions. This is corroborated by own data which proved glial swelling (astrocytes as well as C6 glioma cells) during glutamate exposure [20] in our in vitro model of cytotoxic edema [8–13, 20, 24]. A significant increase of glial cell volume was indeed shown [20] for glutamate in concentrations from 50 μM to 10 mM. Taken together the data underline that glial swelling occurs together with glutamate uptake. Glutamate uptake as well as swelling are only seen if the transmembrane sodium gradient of the glial cells is maintained. It should be mentioned that under pathophysiologic conditions glutamate may reach concentrations in the extracellular space which are well comparable to the highest levels studied. In the vicinity of a traumatic or necrotic focus, where cellular constituents gain access to the interstitial space, extracellular glutamate can be expected to temporarily rise to 5–10 mM. In fact, 1 mM of glutamate has been measured in edematous fluid collected from cat brain with a freezing injury and additional ischemia [1].

Acidosis and Glutamate Uptake

The effect of extracellular acidosis on astrocyte glutamate uptake has recently been studied by Swanson et al. [25]. The authors describe a reduction of glutamate removal from an acidified extracellular environment. This may be due to a reduction of the sodium gradient in deep acidosis due to Na^+/H^+ exchange. A reduced uptake of glutamate and even a reversal of glutamate uptake with a release of glutamate into the extracellular environment has been observed in conditions of deep ischemia [26] when glial homeostasis fails.

Glial Swelling and Cytotoxic Edema

Glial swelling in the course of an uptake process (described above for glutamate) or transport process spatially confined to the vascular environment as postulated for acid extrusion is hence the result of glial homeostasis to protect neurons and their function. This statement, although trivial at first glance, deserves some attention, since glial swelling has long been thought to be an indicator of glial dysfunction, even a first step towards glial decay and cell death. The swelling of cellular elements in the brain is referred to as "cytotoxic brain edema." So far, it has been the general understanding that with impaired energy supply, cell swelling will result from the failure of Na^+/K^+-ATPase according to the pump-leak model of cell volume regulation [16]. Glial swelling in cerebral ischemia occurs fast, within 5 min after interruption of the energy supply. Using our in vitro model again we were able to demonstrate that the breakdown of the energy reserves alone does not suffice to explain the fast swelling response observed in vivo. Our in vitro model was used to evaluate whether anoxia or inhibition of the cellular energy supply without any further changes of the extracellular environment would be accompanied by swelling [9]. Anoxia was induced by discontinuation of the oxygen supply to the membrane oxygenator in an incubation chamber containing the suspended test cells (C6 glioma), and subsequent ventilation with nitrogen. Anoxia was maintained for 2h and at no time was cell swelling observed. The increase in extracellular lactate indicated that anaerobic glycolysis was sufficient to maintain ion gradients and cell volume control. Therefore, iodoacetate was used to inhibit glycolysis in addition to anoxia. This treatment successfully abolished the intra- and extracellular Na^+ gradient within 120 min, but again, cell volume was unaffected [9]. We concluded that in the narrow, restricted extracellular space of the brain in vivo, mediators of secondary brain damage released or activated by pathophysiologic events have to accumulate to cause glial swelling and, eventually, nerve and glial cell death. In another study this conclusion found further support. Here, the sodium potassium pump was blocked by the cardiac glycoside ouabain. The Na^+ gradient was lost by this treatment within 90 min in C6 cells [11]. Simultaneously, a steep drop in the intracellular potassium contents from 190 to 30×10^{-15} mol/cell was observed. Just as found with complete anoxia, the inhibition of the Na^+/K^+-pump did not cause cell swelling [11].

Glial Protection Mechanisms and Reperfusion

Quite evidently the efficacy of glial protection mechanisms is limited under conditions of energy deprivation. Once reperfusion is initiated ion pumps are again supplied with energy and the resulting swelling of glial cells may pose a threat rather than being of any value. In cases of incomplete or low reflow in particular, swelling may even continue: water together with Na^+ ions made available by reperfusion and persisting acidosis will again stimulate swelling. Since the exact activation and inhibition thresholds of the individual transport systems are not known in detail the primary therapeutic goal must be the rapid reestablishment of normal metabolic and hemodynamic conditions. Trials on the efficacy of hypertonic solutions such as those used for small volume resuscitation (4–5 ml/kg bw 7.5% NaCl infused within 3–5 min seem warranted. In order to grant a fast recirculation the inhibition of some glial protection mechanisms might even be useful, e.g., by the use of amiloride analogues in the early reperfusion phase.

Conclusions

The results underline the significance of glial homeostatic mechanisms for neuronal survival during the early phase of ischemia. Glial swelling from various mediators studied so far is, therefore, rather an indicator of glial function than glial demise. In the case of pH regulation, glial swelling and the absence of a normal pH_i considerably enlarge the buffer space and may prevent pH_e from decreasing to levels which go along with immediate neuronal death. pH-regulation in the presence of metabolic energy is carried out by ATP-consuming H^+-ATPases, whereas in ischemic conditions the remaining Na^+ gradient is used to fuel Na^+/H^+-antiport. Glial glutamate carriers prevent uncontrolled interaction of this excitotoxic transmitter with respective receptor sites, again using the sodium gradient as a battery to fuel extracellular homeostasis.

During reperfusion a fast recovery of energy supplies and ionic gradients is essential for a coordinated interaction of homeostatic mechanisms to prevent secondary damage from continuing glial swelling and obstructions of the microcirculation with no-flow areas. The use of hypertonic fluids and antagonists of Na^+/H^+ exchange during early reperfusion might mitigate those dangers by reducing or preventing further swelling.

References

1. Baethmann A, Maier-Hauff K, Schürer L, Lange M, Guggenbichler C, Vogt W, Jakob K, Kempski O (1989) Release of glutamate and free fatty acids in vasogenic brain edema. J Neurosurg 70: 578–591
2. Barbour B, Brew H, Attwell D (1988) Electrogenic glutamate uptake in glial cells is activated by intracellular potassium. Nature 335: 433–435
3. Barbour B, Brew H, Attwell D (1991) Electrogenic uptake of glutamate and aspartate into glial cells isolated from the salamander (Ambystoma) retina. J Physiol (Lond) 436: 169–193
4. Chopp M, Welch KMA, Tidwell CD, Helpern JA (1988) Global cerebral ischemia and intracellular pH during hyperglycemia and hypoglycemia in cats. Stroke 19: 1383–1387
5. Henn FA, Goldstein MN, Hamberger A (1974) Uptake of the neurotransmitter candidate glutamate by glia. Nature 249: 663–664
6. Hertz L, Schousboe A, Boechler N, Mukerji S, Fedoroff S (1978) Kinetic characteristics of the glutamate uptake into normal astrocytes in cultures. Neurochem Res 3: 1–14
7. Katsura K, Ekholm A, Asplund F, Siesjö BK (1991) Extracellular pH in the brain during ischemia: relationship to the severity of lactic acidosis. J Cereb Blood Flow Metab 11: 597–599
8. Kempski O, Chaussy L, Gross U, Zimmer M, Baethmann A (1983) Volume regulation and metabolism of suspended C6 glioma cells. Brain Res 279: 217–228
9. Kempski O, Zimmer M, Neu A, von Rosen F, Baethmann A (1987) Control of glial cell volume in anoxia. Stroke 18: 623–628
10. Kempski O, Staub F, Jansen M, Schödel M, Baethmann A (1988a) Glial swelling during extracellular acidosis in-vitro. Stroke 19: 385–392
11. Kempski O, Staub F, von Rosen F, Zimmer M, Neu A, Baethmann A (1988b) Molecular mechanisms of glial swelling in vitro. Neurochem Pathol 9: 109–125
12. Kempski O, von Rosen F, Weigt H, Staub F, Peters J, Baethmann A (1991) Glial ion transport and volume control. Ann NY Acad Sci 633: 306–317
13. Kempski O, Volk C (1994) Neuro-glial interaction during injury and edema of the CNS. In: Ito U, Hossmann KA, Kuroiwa T, Marmarou A, Reulen HJ, Takakura K (eds) Brain edema IX. Springer Vienna New York, pp 7–11
14. Kraig RP, Chesler M (1988) Glial acid-base homeostasis in brain ischemia. In: Norenberg M, Hertz L, Schousboe (eds) Biochemical pathology of astrocytes. Liss, New York, pp 365–376
15. Kraig RP, Pulsinelli WA, Plum F (1985) Heterogenous distribution of hydrogen and bicarbonate ions during complete brain ischemia. Prog Brain Res 63: 155–165
16. Macknight ADC, Leaf A (1977) Regulation of cellular volume. Physiol Rev 57: 510–573
17. Newman E (1991) Sodium-bicarbonate cotransport in retinal Müller (glial) cells of the salamander. J Neurosci 11: 3972–3983
18. Pappas CA, Ransom BR (1993) A depolarization-stimulated, bafilomycin-inhibitable H^+ pump in hippocampal astrocytes. Glia 9: 280–291
19. Rehncrona S, Siesjö BK, Smith DS (1980) Reversible ischemia of the brain: Biochemical factors influencing restitution. Acta Physiol Scand [Suppl] 492: 135–140

20. Schneider G-H, Baethmann A, Kempski O (1993) Mechanisms of glial swelling induced by glutamate. Can J Physiol Pharmacol 70: S334–S343
21. Schousboe A, Svenneby G, Hertz L (1977) Uptake and metabolism in astrocytes cultured from dissociated mouse brain hemispheres. J Neurochem 29: 999–1005
22. Shrode LD, Putnam RW (1994) Intracellular pH regulation in primary rat astrocytes and C6 glioma cells. Glia 12: 196–210
23. Siesjö BK (1981) Cell damage in the brain: a speculative synthesis. J Cereb Blood Flow Metab 1: 155–185
24. Staub F, Baethmann A, Peters J, Weigt H, Kempski O (1990) Effects of lactacidosis on glial cell volume and viability. J Cereb Blood Flow Metab 10: 866–876
25. Swanson RA, Farrell K, Simon RP (1995) Acidosis causes failure of astrocyte glutamate uptake during hypoxia. J Cereb Blood Flow Metab 15: 417–424
26. Szatkowski M, Barbour B, Attwell D (1990) Non-vesicular release of glutamate from glial cells by reversed electrogenic glutamate uptake. Nature 348: 443–446

On the Role of Glia-Derived Protease Nexin-1 in Neuronal and Glial Adaptation to Ischemia-Induced Cell Loss in Human Brain

C. NITSCH[1], B. EICHE[1,2], J. Ulrich[2], and D. MONARD[3]

Summary

The reexpression of a developmentally regulated serine protease inhibitor with high affinity for thrombin, the glia-derived protease nexin-1, was investigated in thirteen autoptic brains with fresh and old infarcts. In adult intact brain, protease nexin-1 is virtually absent with one exception, the olfactory bulb where degeneration and regeneration persist throughout life. In infarcted brain, in a transition zone between necrotic tissue with glial scar and intact tissue, where degenerating and surviving neuronal elements coexist, protease nexin-1 protein was found in reactive astroglia, provided a delay of 3 days had passed before death. The delay in the onset of the production of the protease inhibitor suggests that, in vivo, it will have only a minor protective effect. In fresh infarcts, macrophages contained phagocytosed protease nexin-1, implying that they participate in the regulation of proteolytic processes. In old malacia, protease nexin-1 was confined to astrocytes in the transition zone where it could be detected even 15 years after the stroke. It is considered that reexpressed protease nexin-1 mimics its developmental role as neurite growth promoting factor by creating a permissive substrate for synaptic rearrangement.

Introduction

Ischemia-induced neuronal degeneration is accompanied by the uncontrolled release of proteases into the extracellular environment. Proteases are set free from degenerating neurons [8, 11] and can reach the neuropil via the opened blood–brain barrier [9, 23, 30], such as the neurotoxic blood-clotting factor thrombin. Uncontrolled spread of the proteases could aggravate and enlarge the primary ischemia-induced lesion. Of course, most proteases have their corresponding inhibitors, and, in case of thrombin, this is protease nexin-1 (PN-1) [1, 29].

PN-1 is a glycoprotein of 43 kDa originally isolated from a glioma cell line on the basis of its neurite growth-promoting activity [16]. Further characterization showed that this glia-derived neurite growth-promoting factor is a potent serine protease inhibitor, belonging to the serpin superfamily [28]. Analysis of the kinetics of PN-1

[1] Institute of Anatomy, University of Basel, Pestalozzistr. 20, 4056 Basle, Switzerland
[2] Institute of Pathology, University of Basel, Schönbeinstrasse 40, 4003 Basle, Switzerland
[3] Friedrich-Miescher Institute, P.O. Box 2543, 4002 Basle, Switzerland

Maturation Phenomenon in Cerebral Ischemia II
U. Ito et al. (Eds.)

interaction with different serine proteases such as thrombin, trypsin, and the plaminogen activators identified thrombin as the preferred target protease, especially in the presence of components of the extracellular matrix [29].

Serine proteases and their inhibitors are particularly highly expressed in the developing nervous system where they cooperate in the regulation of neurite outgrowth [15]. The expression of PN-1 during pre- and postnatal development is widespread and exhibits temporal and regional peaks coinciding with periods of cellular migration and differentiation as shown in mouse [12] and in rat [21]. In the adult brain, selective neurons in layer V of motor and somatosensory cortex, in striatum and in septal nuclei express PN-1 mRNA and contain PN-1 protein [12, 21, 26]. In contrast, glial production and storage of PN-1 has virtually subsided with one exception, the olfactory system [20, 24] where degeneration and regeneration proceed throughout life.

We were recently able to demonstrate that after neuronal injury glial production of PN-1 is upregulated. This is the case in the CA1 sector of the hippocampus after transient global ischemia in the gerbil [7] and the rat [17], and after intracerebral application of glutamate agonists [25]. PN-1 reexpression also occurs in areas with intense terminal degeneration, while nonexcitotoxic loss of perikarya brought about by 6-hydroxydopamine is unable to increase PN-1 content to an immunocytochemically detectable level in astrocytes of the substantia nigra [25]. Thus, an intricate interplay of neurons and glia in which transmitters play an integral part seems to be responsible for the upregulation of PN-1 production in astrocytes of the adult brain. On the other hand, whenever glial reexpression of PN-1 has occurred, its presence is long-lasting, up to 1 year in rat brain [17]. Human PN-1 has been cloned [5, 13] and the protein localized immunohistochemically around blood vessels [3, 31]. Studies on the presence of PN-1 in Alzheimer's disease gave contradictory results [22, 31, 32]. The effect of stroke on the distribution of PN-1 immunoreactivity has not yet been investigated.

Methods

Polyclonal antibodies to human PN-1 were raised in rabbits as described [20]. Their suitability for immunohistochemistry was tested on frozen sections of human olfactory bulbs. Antiserum no. 5 was found to give an excellent immunoreaction for PN-1 in periglomerular astrocytes, closely resembling the situation in rat [20] and gerbil [24]. In paraffinized material, the PN-1 signal is lost (B. Eiche, D. Monard, J. Ulrich, and C. Nitsch, manuscript in preparation).

A first study was carried out in 13 formalin-fixed brains (eight males and five females, aged between 57 and 89 years), obtained in autopsies performed at the Institute of Pathology of the University of Basle, in which the macroscopic aspect revealed the presence of ischemic infarcts. The delay of the cerebrovascular insult ranged between 2 days and 15 years. Not in all cases did the clinical history give the exact date of the stroke and only in a few cases a careful neurological examination had been performed. Thus, the diagnosis was based on the neuropathological findings.

Blocks containing part of the ischemic zone, penumbra and macroscopically intact tissue were postfixed in 4% buffered formaldehyde for 2 days and then submerged for a further 2 days in 30% sucrose for cryoprotection. Alternating 35 μm-thick serial frozen sections were stained with hematoxylin-eosin (HE), cresyl violet and Heidenhain-Woelcke myelin stain, or were subjected to immunoreaction for PN-1 or glial fibrillary acid protein (GFAP) as a marker for reactive astrocytes.

Sections destined for immunohistochemistry were pretreated with 10% normal goat serum. Rabbit anti-PN-1 no. 5 (1:500) and rabbit anti-GFAP (1:800; DAKO Diagnostics, Zug, Switzerland) were diluted in 0.5 *M* Tris-buffered saline (TBS), pH 8.6, containing 0.1% Triton X-100. After incubation for 36 h, biotinylated goat-anti-rabbit IgG (1:200) followed by the avidin-biotin-peroxidase-complex (1:100) were applied twice for 90 min and 60 min each. Visualization of the immunoreaction was done with 0.05% diaminobenzidine and 0.01% H_2O_2 diluted in 0.05 *M* TBS, pH 7.4, containing 1% $NiSO_4$ for enhancement of staining intensity.

To control for false immunopositivity, single steps of the immunoprotocol were omitted; to control for false immunonegativity, olfactory bulbs were coprocessed.

Results

In our hands, with the antiserum used, PN-1 immunoreactivity is virtually absent in intact brain tissue with the exception of the olfactory bulb (B. Eiche, D. Monard, J. Ulrich and C. Nitsch, manuscript in preparation). In particular, no endothelial staining was found. In the neighborhood of the infarcts, however, PN-1 immunoreactivity was present in 12 of the 13 cases investigated. Only in a single case (67-year-old male, with occlusion of the inferoposterior cerebellar artery 2 days before death and fresh malacia of the cerebellar hemisphere) was the PN-1 stain absent although reactive astrocytes were clearly present in the vicinity of the infarct (not shown). In subacute lesions with survival times of 4 days to a few weeks, necrotic tissue was bordered by astroglia and macrophages. Their density decreased when approaching brain tissue with preserved neurons (Fig. 1a). Reactive astrocytes characterized by an overexpression of GFAP were accumulated at the border of the infarct, their number and staining density decreased with increasing integrity of the nervous tissue (Fig. 1b). Exactly in this periinfarct zone, PN-1 immunoreactivity was found in cells exhibiting all the features of astrocytes (Fig. 1c, e). However, in contrast to the GFAP stain which, in varying intensity, was seen all over the brain, PN-1 expression was restricted to a small zone at the border between lesioned and intact tissue. Only in rare instances were PN-1 positive astrocytes situated in the depths of seemingly intact brain tissue (Fig. 1e, f), in particular in patients with multiple infarcts. A unique feature of subacute lesions was the presence of PN-1-containing macrophages and/or microglia intermingled between the PN-1-containing astrocytic processes (Fig. 1d).

In old infarcts, after a delay of months to years, PN-1 containing macrophages had disappeared. Cavities or lacunae were bordered by a glial scar consisting of fibrillary astrocytes devoid of PN-1 immunoreactivity. Beneath the spongiform band the second zone of astrogliosis was PN-1 positive (Fig. 2c). In this area, numerous well-preserved neurons were present (Fig. 2b) but the myelin stain revealed the disturbed neuronal connectivity (Fig. 2a). Old malacia were frequent in our material. It was not possible, however, to determine retrospectively the exact date of the infarct in every case, but in one relatively well documented clinical history the stroke with concurrent contralateral hemiparesis dated back 15 years.

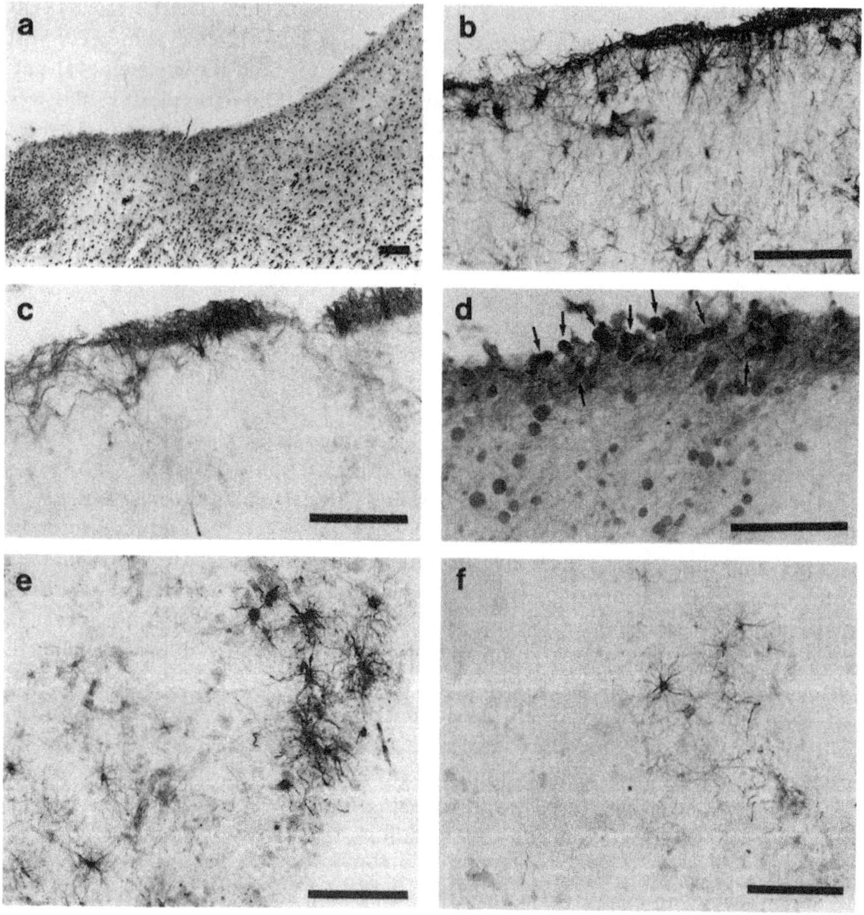

Fig. 1a–f. Multiple subacute lesions in cortex (**a–d**) and basal ganglia (**e,f**) in a 76-year-old male with hypertension and aneurysm of the aorta. *Bar* = 0.1 mm. **a** Loss of nerve cells and accumulation of glial cells, neutrophils and macrophages in a zone bordering the necrotic tissue, the latter lost in part during the staining procedure. H&E stain. **b** intensely glial fibrillary acidic protein (GFAP)-immunoreactive astrocytes in the transition zone between necrotic (lost) and intact brain tissue. A network of astroglial processes demarcates the necrosis. **c** part of the astrocytes in the demarcation zone exhibit protease nexin (PN)-1 immunoreactivity. Note that astrocytes deeper in the tissue are devoid of PN-1; **d** macrophages have invaded the transition zone and many of them possess a perinuclear rim of PN-1 immunoreactivity (*arrows*); **e** PN-1 immunoreactivity in astrocytes and single macrophages in the vicinity of a small anaemic necrosis in the basal ganglia; **f** a group of PN-1 positive astrocytes has no direct contact to an overt lesion

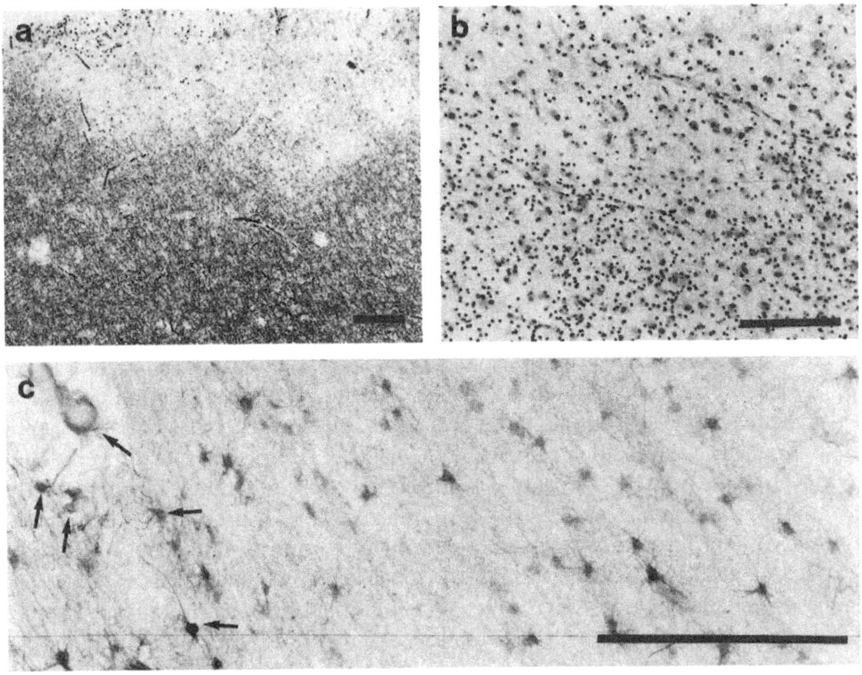

Fig. 2a–c. Old malacia in occipital cortex of an 86-year-old male with arteriosclerotic occlusion of the posterior cerebral artery. *Bar* = 0.2 mm. **a** Resolved infarct reveals in the adjacent tissue the disturbed myeloarchitectonics; Heidenhain-Woelcke stain; **b** Neurons are preserved in this zone, but a slight gliosis is evident; H&E stain. **c** The gliosis in the border zone of this old infarct is accompanied by protease nexin (PN)-1 immunoreactivity in astrocytes. Some of the astrocytes have a perivascular position and extend their processes towards the vessel wall traversing the Virchow-Robin space (*arrows*) but the endothelium proper does not reveal PN-1 immunoreactivity

Discussion

The major findings of the present studies are: (1) The virtual absence of PN-1 immunoreactivity in the adult human brain; (2) the stroke-induced reexpression of PN-1 in astrocytes in the transition zone between infarcted and intact tissue; (3) the presence of PN-1 positive macrophages in subacute lesions; (4) the persistent expression of PN-1 once it is upregulated in adult astrocytes.

1. In contrast to the reports of the Cunningham group [3, 31, 32] we were unable to detect PN-1-positive blood vessels in the intact human brain. This agrees with the observations of Rosenblatt et al. [21] and our own findings in adult rodents where, using a battery of different polyclonal and monoclonal antibodies raised in rabbit, chicken and mouse, with the exception of faint staining of neurons in a few areas [21], glial or endothelial labeling was absent. There is, however, in rodents a discrepancy between the relatively high PN-1 mRNA content and the low PN-1 protein level [12], suggesting that the message is not translated or that the protein is rapidly degraded and/or released. Furthermore, it cannot be excluded that the

protein content is below the detection limit of the assay. Immunohistochemically detectable stored PN-1 protein is found only as a sequel of excitotoxic lesions or nerve terminal degeneration. There it is preferentially localized in astrocytes and these occasionally have a perivascular position [7, 25].

2. Acute brain lesioning is typically accompanied by the appearance of "reactive astrocytes." These particular cells have amplified their GFAP content and can ultimately form a glial scar which is considered to serve as a physical barrier to productive nervous system regeneration [14]. However, not only destructive interventions induce increased GFAP-staining, also subtle alterations in ion homeostasis are a sufficient trigger [10]. Accordingly, reactive gliosis is an early event covering larger areas than those where neuronal degeneration has taken place. Reactive astrocytes can produce and release a variety of factors, including proteases and protease inhibitors [18]. Our data show that after a certain delay, when ischemia-induced neuronal degeneration is apparently at its maximum, reactive astroglia in the transition zone between glial scar and intact tissue start to produce PN-1. In the transition zone, some neurons have survived; others have degenerated and their neuronal connections are in disarray. This is accompanied by considerable alterations in the extracellular cocktail of neurotransmitters and growth factors. In tissue culture, astrocytes are considered to be in a reactive state. Such cultures could be of use to study if some of the multiple neurotransmitter amino acids and growth factors released in the vicinity of the lesion are regulating PN-1 synthesis.

   The delay in the onset of PN-1 production towards the period of irreversible neuronal death suggests that, in vivo, glia-derived PN-1 cannot be considered as an early neuroprotective agent. In fact, in vitro it has a neuroprotective role only when administered before the insult [27]. As PN-1 cannot be detected in the glial scar, it is also unlikely to function as an effective barrier towards extravasated thrombin. It seems more appropriate to consider that glia-derived PN-1 could resume its developmental role in neurite outgrowth and pathfinding in concert with its partner proteases [15] and thus enable synaptic rearrangement in the partially lesioned transition zone. On the basis of this argumentation the occasional patches of PN-1 immunoreactive astrocytes at some distance from the lesion could represent areas of massive terminal degeneration which is accompanied by upregulation of PN-1 synthesis [25].

3. Microglial cells are the resident macrophages of the nervous system. Upon injury they, as well as the circulating macrophages, synthesize and secrete a variety of cytokines and growth factors [4]. Therefore, at first view, it was not surprising to detect PN-1 protein in macrophages in the vicinity of the lesion. Presence of PN-1 has also been shown in microglia of the rat brain in vivo and in vitro but in the absence of PN-1 mRNA [19]. Thus, macrophages are obviously able to internalize native or complexed astrocyte-derived PN-1 and, by removing it from the extracellular environment, to participate in the regulation of proteolytic processes during synaptic rearrangement.

4. The persistent upregulation and/or storage of PN-1 in adult astrocytes once this has been induced by excitotoxic injury in the adult brain can in part be explained by the fact that ischemia-induced degeneration is not a transitory phenomenon occurring only during a narrow time-window, but takes a prolonged time course

[2]. Transneuronal anterograde and retrograde degeneration proceeding after focal injury in more or less distant brain areas and lasting for months in rats [6] may also necessitate a continuous synaptic rearrangement. The question, however, remains whether the presence of a protease inhibitor normally only produced in high concentrations during development may have positive or adverse effects on the remaining tissue.

In conclusion, PN-1 protein, which is normally below detection level in adult human brain, is produced in astrocytes after an ischemic insult. However, astrocytes in the glial scar do not contain PN-1. The fact that PN-1 is only expressed in a transition zone where degenerating and intact neuronal elements coexist suggests that it assists in creating a permissive environment for synaptic rearrangement.

References

1. Baker JB, Low DA, Simmer RL, Cunningham DD (1980) Protease-nexin: a cellular component that links thrombin and plasminogen activator and mediates their binding to cells. Cell 21: 37–45
2. Bonnekoh P, Barbier A, Oschlies U, Hossmann K-A (1990) Selective vulnerability in the gerbil hippocampus 8 morphological changes after 5-min ischemia and long survival times. Acta Neuropathol (Berl) 80: 18–25
3. Choi BH, Suzuki M, Kim T, Wagner SL, Cunningham DD (1990) Protease nexin 1. Localization in the human brain suggests a protective role against extravasated serine proteases. Am J Pathol 137: 741–747
4. Dickson WD, Lee SC, Mattiace LA, Yen S-HC, Brosnan C (1993) Microglia and cytokines in neurological disease, with special reference to AIDS and Alzheimer's disease. Glia 7: 75–83
5. Gloor S, Odink K, Guenther J, Nick H, Monard D (1986) A glia-derived neurite promoting factor with protease inhibitor activity belongs to the protease nexins. Cell 47: 687–693
6. Hara H, Harada K, Sukamoto T (1993) Chronological atrophy after transient middle cerebral artery occlusion in rats. Brain Res 618: 251–260
7. Hoffmann M-C, Nitsch C, Scotti AL, Reinhard E, Monard D (1992) The prolonged presence of glia-derived nexin, an endogenous protease inhibitor, in the hippocampus after ischemia-induced delayed neuronal death. Neuroscience 49: 397–408
8. Hong S-C, Lanzino G, Goto Y, Kang SK, Schottler F, Kassell NF, Lee KS (1994) Calcium-activated proteolysis in rat neocortex induced by transient focal ischemia. Brain Res 661: 43–50
9. Ito K, Go KG, Walker JT Jr., Spatz M, Klatzo I (1976) Experimental cerebral ischemia in Mongolian gerbils. III. Behaviour of the blood brain barrier. Acta Neuropathol (Berl) 34: 1–6
10. Kraig RP, Dong L, Thisted R, Jaeger CB (1991) Spreading depression increases immunohistochemical staining of glial fibrillary acidic protein. J Neurosci 11: 2187–2198
11. Lee KS, Frank S, Vanderklish P, Arai A, Lynch G (1991) Inhibition of proteolysis protects hippocampal neurons from ischemia. Proc Natl Acad Sci USA 88: 7233–7237
12. Mansuy IM, van der Putten H, Schmid P, Meins M, Botteri FM, Monard D (1993) Variable and multiple expression of protease nexin-1 during mouse organogenesis and nervous system development. Development 119: 1119–1134
13. McGrogan M, Kennedy J, Li MP, Hsu C, Scott RW, Simonsen CC, Baker JB (1988) Molecular cloning and expression of two forms of human protease nexin-1. Biotechnology 6: 172–177
14. McKeon RJ, Schreiber RC, Rudge JS, Silver J (1991) Reduction of neurite outgrowth in a model of glial scarring following CNS injury is correlated with the expression of inhibitory molecules on reactive astrocytes. J Neurosci 11: 3398–3411
15. Monard D (1988) Cell-derived proteases and protease inhibitors as regulators of neurite outgrowth. Trends Neurosci. 11(12): 541–544
16. Monard D, Solomon F, Rentsch M, Gysin R (1973) Glia-induced morphological differentiation in neuroblastoma cells. Proc Natl Acad Sci USA 70: 1894–1897

17. Nitsch C, Scotti AL, Monard D, Heim C, Sontag KH (1993) The glia-derived protease nexin 1 persists for over 1 year in rat brain areas selectively lesioned by transient global ischaemia. Eur J Neurosci 5: 292–297
18. Norenberg MD (1994) Astrocyte responses to CNS injury. J Neuropathol Exp Neurol 53: 213–220
19. Pavlik A, Monard D •• (submitted)
20. Reinhard E, Meier R, Halfter W, Rovelli G, Monard D (1988) Detection of glia-derived nexin in the olfactory system of the rat. Neuron 1: 387–394
21. Reinhard E, Suidan HS, Pavlik A, Monard D (1994) Glia-derived nexin/protease nexin-1 is expressed by a subset of neurons in the rat brain. J Neurosci Res 37: 256–270
22. Rosenblatt DE, Geula C, Mesulam M-M (1989) Protease nexin 1 immunostaining in Alzheimer's disease. Ann Neurol 26: 628–634
23. Schmidt-Kastner R, Szymas J, Hossmann KA (1990) Immunohistochemical study of glial reaction and serum-protein extravasation in relation to neuronal damage in rat hippocampus after ischemia. Neuroscience 38: 527–540
24. Scotti AL, Hoffmann M-C, Nitsch C (1994) The neurite growth promoting protease nexin 1 in glial cells of the olfactory bulb of the gerbil: an ultrastructural study. Cell Tissue Res 278: 409–413
25. Scotti AL, Monard D, Nitsch C (1994) Re-expression of glia-derived nexin/protease nexin 1 depends on mode of lesion-induction or terminal degeneration: observations after excitotoxin or 6-hydroxydopamine lesions of rat substantia nigra. J Neurosci Res 37: 155–168
26. Simpson CS, Johnston HM, Morris BJ (1994) Neuronal expression of protease-nexin 1 mRNA in rat brain. Neurosci Lett 170: 286–290
27. Smith-Swintosky VL, Zimmer S, Fenton JW II, Mattson MP (1995) Protease nexin-1 and thrombin modulate neuronal Ca^{2+} homeostasis and sensitivity to glucose deprivation-induced injury. J Neurosci 15: 5840–5850
28. Sommer J, Gloor SM, Rovelli GF, Hofsteenge J, Nick H, Meier R, Monard D (1987) cDNA sequence coding for a rat glia-derived nexin and its homology to members of the serpin superfamily. Biochemistry 26: 6407–6410
29. Stone SR, Nick H, Hofsteenge J, Monard D (1987) Glia-derived neurite-promoting factor is a slow-binding inhibitor of trypsin, thrombin, and urokinase. Arch Biochem Biophys 252: 237–244
30. Suzuki R, Yamaguchi T, Kirino T, Orci F, Klatzo I (1983) The effects of 5-minute ischemia in Mongolian gerbils. I. Blood-brain barrier, cerebral blood flow and local cerebral glucose utilization changes. Acta Neuropathol (Berl) 60: 207–216
31. Vaughan PJ, Su J, Cotman CW, Cunningham DD (1994) Protease nexin-1, a potent thrombin inhibitor, is reduced around cerebral blood vessels in Alzheimer's disease. Brain Res 668: 160–170
32. Wagner SL, Geddes JW, Cotman CW, Lau AL, Isackson PJ, Cunningham DD (1989) Protease nexin-1, a protease inhibitor with neurite promoting activity, is reduced in Alzheimer's disease. Proc Natl Acad Sci USA 86: 8284–8288

Functional Outcome After Brain Infarction: Effect of Enriched Environment and Amphetamine

B.B. Johansson, B. Mattsson, and A.L. Ohlsson

Summary

Rats housed in an enriched environment after focal brain ischemia improved more than rats housed under standard laboratory conditions. The aim was to investigate if transfer to an enriched environment 1 h per day had the same favorable effect as permanent housing enrichment, and if amphetamine further improved outcome in rats housed in an enriched environment. After distal ligation of the right middle cerebral artery, rats were divided into four groups: permanent enriched environment with and without amphetamine or standard laboratory cages with a 1 h day transfer to enriched environment with and without amphetamine. *d,l*-Amphetamine 2 mg/kg or vehicle was given on postoperative day 5–7 and repeated 2, 4 and 7 days later. Leg placement, sensory motor integration, paw reaching, balance on an inclined plane or when traversing a rotating pole were tested 1–8 weeks postoperatively. Infarct volume or thalamic atrophy did not differ between the groups. Rats exposed to environmental enrichment 1 h a day improved to about the same extent as rats in permanent environmental enrichment. Amphetamine did not enhance the outcome in any of the tests.

Introduction

In contrast to the intense research on how to rescue ischemic neurons in the acute stage of brain ischemia, the question of why functions can improve with time in spite of permanent tissue damage has attracted little interest. Various possible mechanisms behind functional improvement after stroke have been discussed, including resolution of brain edema, absorption of necrotic tissue and disappearance of remote functional depression or "diaschisis." Several factors might be involved but current data suggest that the adult brain retains a capacity for plasticity and functional reorganization and that a substantial part of functional recovery after stroke might be attributed to brain plasticity [1, 15, 17, 25–28].

In agreement with studies on other types of acute brain lesions [30] we have shown that rats housed in enriched environment perform significantly better than rats housed in standard cages after a focal brain infarction [11, 16, 18, 21]. There is also evidence that drugs can influence the recovery process [5–9, 13, 14, 29]. Catechola-

Department of Neurology, Lund University Hospital, 22185 Lund, Sweden

Maturation Phenomenon in Cerebral Ischemia II
U. Ito et al. (Eds.)

mines are thought to be involved in learning and memory and several studies have reported that, at least if combined with test specific training, norepinephrine, amphetamine and other α-adrenergic stimulating drugs can enhance recovery.

The aim of the present study was to answer the following two questions: (1) Is it necessary to keep the rats in a permanent enriched environment or can a limited daily exposure have a similar effect? (2) Can amphetamine further improve functional outcome also in rats housed in enriched environment?

Material and Methods

Four-month-old male, spontaneously hypertensive rats were anesthetized with methohexital sodium and the right middle cerebral artery (MCA) was occluded distal to the origin of the striatal branches. Twenty-four hours following the operation the rats were housed in individual cages. Thereafter, the rats were divided into four groups. Group A and group B were permanently kept in an enriched environment; group A ($n = 12$) without and group B ($n = 12$) with the addition of amphetamine. Two groups were kept in standard laboratory cages with a maximum of five rats in each cage and transferred to an enriched environment 1 h per day starting at the time of the first amphetamine administration; group C ($n = 9$) with no other treatment and group D ($n = 10$), which received amphetamine. On postoperative day 5–7 and repeated 2, 4 and 7 days later, vehicle or 2 mg/kg *d,l*-amphetamine was administered i.p. early in the evening when rats spontaneously start to become active, in groups C and D directly following the 1-h exposure to enriched environment. The enriched environment box has been described earlier [11, 21]. In short, ten rats are housed together in a large furnished cage that allowed various activities but the animals are not forced to do any specific exercises.

Leg placement, sensory motor integration, paw reaching, balance on an inclined plane or when traversing a rotating pole were tested 1–8 weeks postoperatively. In the *limb placing test*, shortened and modified after De Ryck [4] as earlier described [21], the forelimb and hindlimb placements were evaluated by an observer with no knowledge of which group the rat belonged to. During the *sensorimotor integration test*, the animals were placed on an elevated 150-mm wide circular platform [10]. The fur was then touched with a pointed wooden pen at seven regions on each side of the body. For each region the extent of orientation was rated according to the following scale: 0, no head orientation toward touch; 1, slight head turning toward touch; 2, turning the head approximately half-way in the direction of touch; 3, stretching by the hind paw or precise localization of the head to body region touched; 4, precise localization of the head and biting of pen. Touching of the paw could also result in a slow or fast removal of the limb, which was rated as a score of 1 or 2, respectively. The regional scores were added separately for each side. The more alert the rat, the higher the scores on both sides. The variation between individual rats was compensated for by giving the data for the difference between the scores on the ipsilateral healthy side and the contralateral side.

Paw-reaching was evaluated with a staircase test [10, 20] in a dark room. Custom-made plexiglass boxes with a central elevated platform with a staircase with six steps are used. Each step was baited with three 45-mg chow pellets (Cambden Instruments Ltd., Loughborough, UK). The rat was placed on the platform for 5 min and the number of pellets taken by the left and right paw counted. The rats were tested over 4 preoperative days and over 5 days 3 and 7 weeks after the MCA occlusion. During the last 2 days of the last test period, the staircase was only baited on the left side. The test was preceded by a 48-h period of food deprivation. After each test, the rat was given about 20 g of standard food pellets to maintain body weight. The *inclined plane* was constructed of two rectangular plywood boards connected by a hinge allowing adjustment of the inclination [24]. A rubber mat was fixed to the surface of the movable board. The maximum angle of the plane at which the animal could maintain its position without falling was recorded. Coordination and integration of motor movement was tested on a *rotating pole*. The pole, 45 mm in diameter and 1500 mm in length, rotated alternately to the left or the right with three or ten turns per min and the behavior was scored from 0 at which the rat falls down to 6 at which the rat crosses the pole with no foot slips [21].

Determination of Infarct Volume

Nine weeks after the MCA ligation, the rats were deeply anesthetized with methohexital and the brains were removed and frozen in isopentane chilled to −40°. The brains were stored at −80°C until 20-μm thick coronal sections were cut in a cryostat at −20°. At 300-μm intervals, sections were taken for determination of infarct volume and thalamic atrophy, a total of 15 sections from each rat [21]. The "total infarct volume" was calculated according to percentage of total hemispheric volume (the size of the contralateral hemisphere minus the remaining tissue on the lesioned side) and the cortical part of the infarct presented in terms of percentage of total cortex. The "total infarct volume" will at this late stage after an MCA occlusion be overestimated and include secondary tissue loss, particularly in the thalamus. The thalamic volume was therefore determined separately. The image analyzing system consisted of a video camera (Dage MTI, Midilgan City, IN, USA), a light box (Imaging Research Inc., St. Catherines Ontario, Canada) and a Macintosh computer as described earlier.

Statistics

For tests based on scoring systems (ordinal measures) the Kruskal Wallis non-parametric analysis of variance test was used with a multiple comparison post hoc test to determine the number and relation of the group differences at 95% significance level. For difference in infarct volume and behavioral tests with interval scales, one-way parametric analysis of variance (ANOVA) with Scheffe's post hoc procedure was used at 95% significance levels.

Results

There was no major difference in behavior between the groups. At no time was there any group difference in the limb placement tests (Table 1). There was a trend towards better performance in the sensorimotor integration test in rats given amphetamine but with a significant difference only between groups B and C at 8 weeks, that is, between the group with limited exposure to enriched environment and the permanently enriched amphetamine group (Fig. 1). On the rotating pole and inclined plane there was a slight trend in that rats with limited exposure to enriched environment performed more poorly than those with permanent enrichment (Figs. 2, 3). When only the left side of the staircase was baited during the last 2 days of the paw reaching test, the number of pellets consumed increased markedly in rats with limited enriched environment and no amphetamine, whereas the effect in the other groups was rather small (Fig. 4). Infarct and thalamic volume did not differ between the groups (Table 2).

Table 1. Mean neurological scores for the affected left paws in a limb placement test after ligation of the right middle cerebral artery (MCA)

Time after MCA occlusion	Groups			
	A ($n = 12$)	B ($n = 12$)	C ($n = 9$)	D ($n = 10$)
1 week	5.6 ± 2.4	5.7 ± 2.6	6.4 ± 3.2	4.9 ± 4.0
5 weeks	5.9 ± 1.8	6.4 ± 2.6	7.3 ± 2.9	7.2 ± 2.9
6 weeks	8.3 ± 3.0	8.0 ± 2.1	9.1 ± 3.5	9.4 ± 1.9
8 weeks	8.4 ± 3.9	8.6 ± 3.8	7.7 ± 4.3	9.2 ± 2.4

Mean ± SD. Score 16 = normal behavior.
A, Rats housed in enriched environment after MCA occlusion; B, rats housed in enriched environment after MCA occlusion; amphetamine; C, rats kept in standard laboratory cages of five but exposed to enriched environment 1 h a day; D, rats kept in standard laboratory cages of five but exposed to enriched environment 1 h a day; amphetamine.

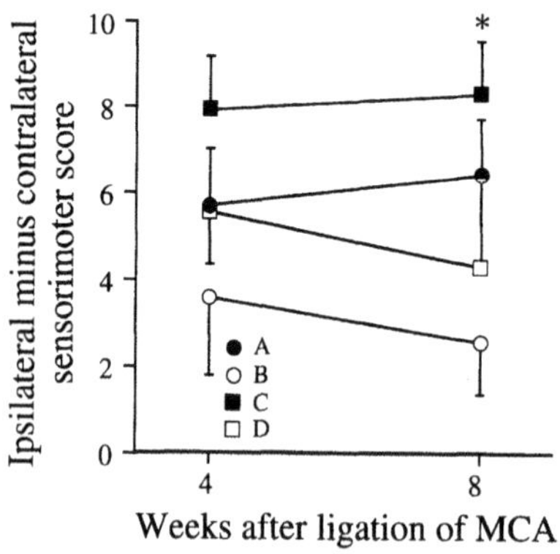

Fig. 1. Sensorimotor integration scores expressed as the difference between the ipsilateral unaffected side and the impaired side. Thus the best performance gives the lowest score. Groups *A* and *B*, housed in permanent enriched environment; *B* + amphetamine; Groups *C* and *D*, exposed to enriched environment 1 h a day; *D* + amphetamine. *Significant difference between groups *B* and *C*; *MCA*, middle cerebral artery

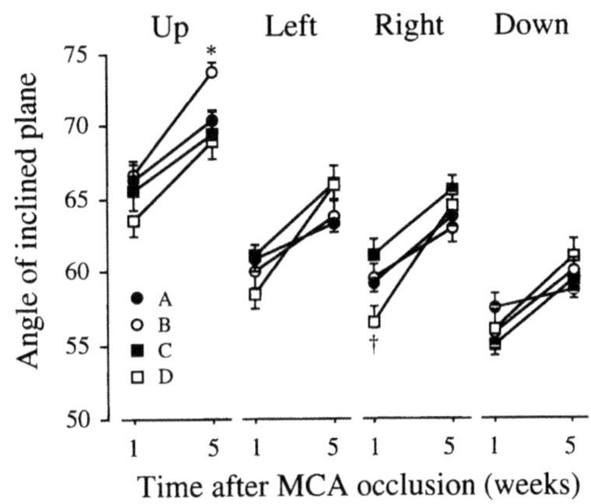

Fig. 2. The maximum angle of the inclined plane at which the rats could maintain their position with the head facing up, left, right or down. *Significant difference between *B* and *D*; † significant difference between *C* and *D*. *MCA*, middle cerebral artery

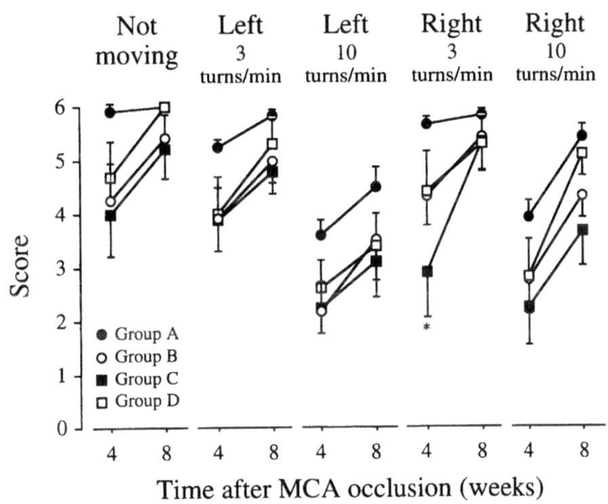

Fig. 3. Performance score on a rotating pole 4 and 8 weeks after the middle cerebral artery (*MCA*) occlusion. *Score 6* indicates that the rat crosses the pole without any problem. **Group C* significantly different from *group A*

Discussion

The beneficial effect of amphetamine has been proposed as being due to stimulation of functionally depressed catecholamine systems. *d,l*-Amphetamine stimulates dopaminergic as well as adrenergic receptors, whereas *d*-amphetamine stimulates predominantly dopamine receptors. In the early study on the capacity of amphetamine to facilitate recovery after focal brain injury, haloperidol blocked the amphetamine-promoted restoration of motor function in rats and it impaired recovery when

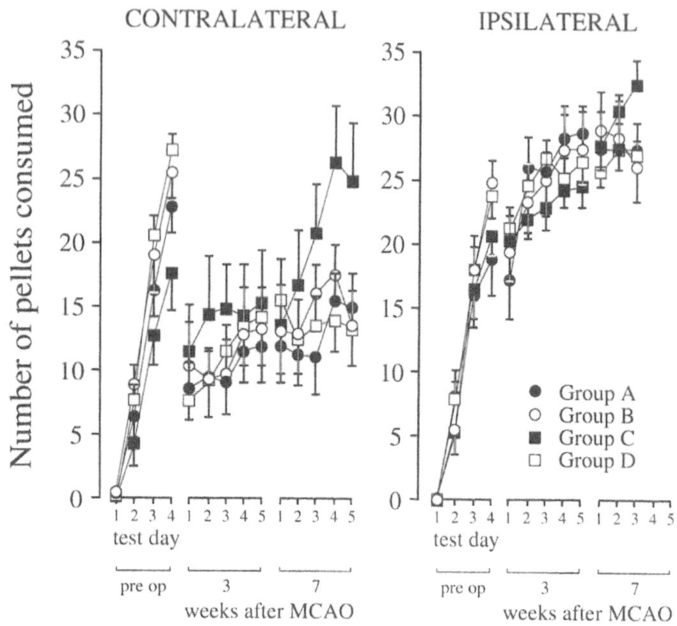

Fig. 4. Number of pellets consumed in the paw reaching test preoperatively and 3 and 7 weeks after occlusion of the right middle cerebral artery (*MCAO*)

Table 2. Infarct and thalamic volumes 9 weeks after ligation of the middle cerebral artery (MCA)

Group	No.	Total infarct (% of contralateral hemisphere)	Cortical infarct (% of contralateral cortex)	Thalamus (% of contralateral thalamus)
A	12	16.1 ± 4.9	25.4 ± 5.4	80.7 ± 11.9
B	12	16.7 ± 10.1	26.5 ± 12.2	81.2 ± 13.1
C	9	12.9 ± 6.9	21.8 ± 11.2	80.0 ± 8.4
D	10	14.5 ± 5.1	25.3 ± 8.4	85.0 ± 6.4

Mean ± SD.
A, Rats housed in enriched environment after MCA occlusion; B, rats housed in enriched environment after MCA occlusion; amphetamine; C, rats kept in standard laboratory cages of five but exposed to enriched environment 1 h a day; D, rats kept in standard laboratory cages of five but exposed to enriched environment 1 h a day; amphetamine.

given alone [5]. Several later studies have shown that norepinephrine and some other α-adrenergic stimulating drugs have a similar effect to amphetamine [6, 7, 29].

Norepinephrine is thought to play an important role in cortical plasticity. Several studies have shown that neonatal norepinephrine, but not dopamine, depletion abolishes or attenuates the effect of enriched environment on neural and behavioral measures [19, 22, 23]. However, results of some recent studies indicate that the effect or norepinephrine depletion may at least in part be transient or can be compensated for. Thus, xylamine (*N*-2-chloroethyl-2-methylbenzylamine), a noradrenergic neurotoxin, prevented the increase in weight in cortex seen in nontreated animals during maze training, but did not impair the spatial working memory and investigatory behavior of the animals [2]. In another study, depletion of norepinephrine in adult rats did not impair their general exploration but the social memory was transiently reduced. However, complete functional recovery occurred with time in the presence of vitually total brain noradrenergic depletion [12].

It is usually stressed that amphetamine and norepinephrine have to be combined with task-specific training in order to enhance recovery [5, 6, 9, 29], but that does not always seems to be necessary. In a study in cats, single does of *d*-amphetamine given 10 days after unilateral frontal cortex ablation produced an immediate and enduring acceleration of beam-walking ability if the cats were exposed to beam-walking training while under amphetamine intoxication. Four doses of amphetamine at 4-day intervals promoted recover faster than a single dose, and rats with no beam walking experience during the intoxication but tested on the beam before the amphetamine administation did as well at the end of the treatment, even if the initial recovery time course was slower. The authors commented that multiple doses of the drug apparently override the necessity for task-relevant experience, possibly due to generalization from nonspecific locomotor experience [13]. Likewise, in a study of rats pretrained in a T-maze performance, *d*-amphetamine-treated rats improved faster and more completely after thrombotic cortical infarction than saline injected rats within the postoperative time studied although amphetamine was administered 24h before the T-maze training [14].

The fact that we found no additional benefit of amphetamine when given to rats exposed to environmental enrichments may indicate that amphetamine and environmental enrichment influence the brain in a common way. One nonspecific mechanism could be arousal. It has been shown that adrenal demedullation attenuates the amphetamine-induced enhancement of learning [31]. The fact that amphetamine had no effect on forelimb tactile placing in the cat [13], a deficit that is significantly reduced in rats exposed to enriched environment [16, 21], may indicate different mechanisms. A direct comparison, however, with amphetamine administration to rats housed in standard laboratory environment with rats environmentally enriched would be needed to determine if there is a difference in specific tests. Environmental enrichment can markedly alter brain morphology and biochemistry [17, 26, 30] and it is perhaps more likely that the effect that can be obtained in an enriched environment is more extensive but includes effects that can also be obtained with amphetamine.

Acknowledgments. Works from the author's own laboratory was supported by grants from the Swedish Medical Research Council (Project 14X-4968), the Bank of Sweden Tercentenary Foundation, the 1987 Stroke Foundation, and King Gustaf V's and Queen Victoria's Foundation.

References

1. Bach-y-Rita P (1990) Brain plasticity as a basis for recovery of function in humans. Neuropsychologica 28: 547–554
2. Benloucif S, Bennett EL, Rosenzweig MR (1995) Norepinephrine and neural plasticity: the effects of xylamine on experience-induced changes in brain weight, memory, and behavior. Neurobiol Learn Mem 63: 33–42

3. Coyle P (1992) Middle ceebral artery occlusion in the young rat. Stroke 13: 855–859
4. De Ryck M, Van Reempts J, Borgers M, Wauquier A, Janssen AJ (1989) Photochemical stroke model: flunarizine prevents sensorimotor deficits after neocortical infarcts in rats. Stroke 20: 1383–1390
5. Feeney DM, Gonzalez A, Law WA (1982) Amphetamine, haloperidol and experience interact to affect the rate of recovery after motor cortex injuries. Science 217: 855–857
6. Feeney DM, Weisend MP, Kline AE (1993) Noradrenergic pharmacotherapy, intracerebral infusion and adrenal transplantation promote functional recovery after cortical damage. J Neurol Transplant Plast 4: 199–213
7. Goldstein LB (1993) Basic and clinical studies on pharmacologic effects on recovery from brain injury. J Neurol Transplant Plast 4: 175–192
8. Goldstein LB (1995) Common drugs may influence motor recovery after stroke. Neurology 45: 865–871
9. Goldstein LB, Davis JN (1990) Post-lesion practice and amphetamine-facilitated recovery of beam-walking in the rat. Restor Neurol Neurosci 1: 311–314
10. Grabowski M, Brundin P, Johansson BB (1993) Paw-reaching, sencorimotor, and rotational behavior after brain infarction in rats. Stroke 24: 899–902
11. Grabowski M, Sørensen JC, Mattsson B, Zimmer J, Johansson BB (1995) Influence of an enriched environment and cortical grafting on functional outcome in brain infarcts of adult rats. Exp Neurol 133: 96–102
12. Griffin MG, Taylor GT (1995) Norepinephrine modulation of social memory: evidence for a time-dependent functional recovery of behavior. Behav Neurosci 109: 466–473
13. Hovda DA, Feeney DM (1984) Amphetamine with experience promotes locomotor function after unilateral frontal cortex injury in the cat. Brain Res 298: 358–362
14. Hurwitz BE, Dietrich WD, McCabe PM, Alonson O, Watson BD (1991) Amphetamine promotes recovery from sensory-motor integration deficit after thrombotic infarction of the primary somatosensory rat cortex. Stroke 22: 648–654
15. Jenkins WM, Merzenich MM (1987) Reorganization of neocortical representations after brain injury: a neurophysiological model of the bases of recovery from stroke. Prog Brain Res 71: 249–266
16. Johansson BB (1996) Functional outcome in rats transferred to an enriched environment 15 days after focal brain ischemia. Stroke 27: 324–326
17. Johansson BB, Grabowski M (1994) Functional recovery after brain infarction: plasticity and neural transplantation. Brain Pathol 4: 85–95
18. Johansson BB, Ohlsson A-L (1996) Environment, social interaction and physical activity as determinats of functional outcome after cerebral infarction in the rat. Exp Neurol 139: 322–327
19. Mohammed AK, Jonsson G, Archer T (1986) Selective lesioning of forebrain noradrenaline neurons at birth abolishes the improved maze learning performance induced by rearing in complex environment. Brain Res 398: 6–10
20. Montoya CP, Campbell-Hope LJ, Pemberton KD, Dunnett SB (1991) The "staircase test": a measure of independent forelimb reaching and grasping abilities in rats. J Neurosci Methods 36: 219–228
21. Ohlsson A-L, Johansson BB (1995) The environment influences functional outcome of cerebral infarction in rats. Stroke 26: 644–649
22. O'Shea L, Saari M, Pappas BA, Ings R, Stange K (1983) Neonatal 6-hydroxydopamine attenuates the neuronal and behavioral effect of enriched rearing in the rat. Eur J Pharmacol 92: 43–47
23. Pappas BA, Saari M, Smythe J, Murtha S, Stange K, Ings R (1987) Forebrain norepinephrine and neurobehavioral plasticity: neonatal 6-hydroxydopamine eliminates enriched-impoverished experience effects on maze performance. Pharmacol Biochem Behav 27: 153–158
24. Rivlin AS, Tator CH (1977) Objective clinical assessment of motor function after experimental spinal cord injury in the rat. J Neurosurg 47: 577–581
25. Pons TP, Garraghty PE, Mishkin M (1988) Lesion-induced plasticity in the somatosensory cortex of adult macaques. Neurobiology 85: 5279–5281
26. Rosenzweig MR (1984) Experience, memory and the brain. Am Psychol 4: 365–376
27. Schieber MH (1995) Physiological bases for functional recovery. J Neuro Rehabil 9: 65–71
28. Stein DG, Glasier MM (1992) An overview of developments in research on recovery from brain injury. In: Rose FD, Johnson DA (eds) Recovery from brain damage. Plenum, New York, pp 1–22

29. Sutton RL, Feeney DM (1992) a-noradrenergic agonists and antagonists affect recovery and maintenance of beam-walking ability after senorimotor cortex ablation in the rat. Rest Neurol Neurosci 4: 1–11
30. Will B, Kelche C (1992) Environmental approaches to recovery of function from brain damage: a review of animal studies (1981–1991). In: Rose FD, Johnson DA (eds) Recovery from brain damage. Plenum, New York
31. Walsh R (1981) Sensory environments, brain damage, and drugs: a review of interactions and mediating mechanisms. Int J Neurosci 14: 129–137

Behavioral Changes After Cerebral Infarction in the Rat: Long-Term Spatial Cognitive Impairment and Cerebral Acetylcholine Levels

A. Tamura[1], H. Nagashima[1], M. Hirakawa[1], K. Narita[1], Y. Nagashima[1], H. Nakayama[1], F. Yonemori[2], M. Okada[3], M. Fujiwara[3], and K. Sano[1]

Summary

The aim of the present study was to observe behavioral changes following middle cerebral artery (MCA) occlusion in the rat using the Morris water maze task and the radial eight-arm maze task, and to investigate the relationship between alterations in regional levels of acetylcholine (ACh) in the brain and disturbance of spatial cognition. When the Morris water maze task sessions were started at 8 weeks after MCA occlusion, the escape latency and swimming path length were significantly increased in the MCA-occluded rats as compared with sham-operated rats. When the MCA was occluded after a preoperative acquisition trial of this task, retention trials 8 weeks after MCA occlusion showed significant increases in the escape latency and path length. In the study using the radial 8-arm maze task, about two-thirds of the MCA-occluded rats showed long-term spatial cognitive impairment, while one-third of them were able to perform this task. ACh levels in the infarcted areas decreased considerably in all of the MCA-occluded rats. However, ACh levels increased significantly in the parietal cortex in rats which could perform the task. These results suggest that the cholinergic function of the parietal cortex might play a role in spatial cognition in the MCA-occluded rat.

Introduction

Unilateral middle cerebral artery (MCA) occlusion in the rat developed by Tamura and coworkers [10] is widely used as a model of focal cerebral ischemia, and numerous studies have been performed on MCA-occluded rats using biological, pathological and electrophysiological approaches. Experimental studies of cerebrovascular diseases using MCA-occluded rats have mainly focused on changes in cerebral blood flow, cerebral metabolism and neuropathological findings [2, 11, 12]. With a view to improving the treatment of stroke patients, memory and learning ability after stroke is important. However, there are few reports concerning behavioral changes after

[1] Department of Neurosurgery, Teikyo University School of Medicine, 2-11-1, Kaga, Itabashi-ku, 173, Tokyo, Japan
[2] Central Pharmaceutical Research Institute, Japan Tobacco Inc., 1-1 Murasaki-cho, Takatsuki, Osaka 569, Japan
[3] Department of Physiology and Pharmacology, Faculty of Pharmaceutical Sciences, Fukuoka University, 8-19-1, Nanakuma, Jonan-ku, Fukuoka 814-01, Japan

Maturation Phenomenon in Cerebral Ischemia II
U. Ito et al. (Eds.)

focal cerebral ischemia in animals. Therefore, we studied behavioral changes in MCA-occluded rats. For evaluation of behavioral changes in small animals such as rats and mice, the following methods have been used: (1) the passive avoidance task, (2) the active avoidance task, and (3) the maze test. In 1988, we reported disturbance of memory acquisition when training for the one-trial passive avoidance task was conducted 3 days after MCA occlusion and the test trial was started on postischemic day 4 and then on day 14 [15]. In subsequent studies, several authors have reported observing behavioral changes after MCA occlusion in rat using a passive avoidance response or maze task [3, 5, 7, 8, 13]. We have also reported that behavioral changes in MCA-occluded rats are ameliorated by treatment with a thyrotropine-releasing hormone derivative without any effect on the severity of ischemic brain damage [16, 17]. In most of these studies animals were trained postoperatively for the purpose of observing their memory acquisition. Then, we investigated memory retention ability in cases in which training of a one-trial passive avoidance task or active avoidance task was conducted before MCA occlusion, and reported that retention of memory in both tasks was disturbed after MCA occlusion in the rat [3].

The aim of the present study was to observe behavioral changes over a long period in the chronic phase of focal cerebral ischemia, while focusing on deficits in spatial cognition following MCA occlusion in the rat using the Morris water maze task and the radial eight-arm maze task and to investigate the relationship between alterations in regional brain levels of acetylcholine (ACh) and disturbances in spatial cognition.

Surgical Preparation

Male Wistar rats were anesthetized with 2% halothane and the proximal portion of the MCA was permanently occluded using a microsurgical technique which was modified from the original technique developed by Tamura et al. for the purpose of performing chronic experiments [10, 15]. The stem of the MCA was electrocauterized medial to the olfactory tract and was cut to ensure complete vascular occlusion. In the sham-operated group, the MCA was only exposed without occlusion.

Water Maze Task

The spatial learning and memory of the rats was examined using the Morris water maze task [6, 18]. The apparatus used was a cylindrical tank filled with water. A transparent platform was placed in a constant position in the center of one quadrant within the tank. The platform was set 2 cm below the water level where the rats could not see it directly. In each trial, rats were placed into the water at a fixed starting position and the time taken to escape onto the hidden platform and the swimming path length were measured using a video image motion analyzer.

In the acquisition study, acquisition trials were carried out for 10 successive days 8 weeks after MCA occlusion. The sham-operated rats ($n = 11$) performed this task rapidly; in the last session, they took 14.0 ± 12.8 s. MCA-occluded rats ($n = 11$) swam well and adopted a searching strategy. In the first session, the swimming path length of MCA-occluded rats tended to be longer than that of sham-operated rats, but the difference was not significant. However, the MCA-occluded rats did not reach normal performance levels after 10 days of acquisition trials. The mean time taken to complete the task in this group in the last session was 66.3 ± 31.9 s. The escape latency and the swimming path length were significantly longer than those of the sham-operated rats during the last three session ($p < 0.01$). In contrast to the escape latency and swimming path length, there was no significant difference with respect to swimming speed.

In a retention study, acquisition trials were carried out for 14 successive days prior to MCA surgery. The postoperative retention trial was carried out for 3 days 8 weeks after MCA occlusion. Both the sham-operated group ($n = 12$) and the MCA-occluded group ($n = 18$) performed the task in the acquisition trial; in the last session, the sham-operated group took 4.5 ± 2.1 s (mean ± SD) to reach the platform, and the MCA-occluded group took 4.6 ± 1.7 s. There were no significant differences between the two groups. In the first session of the postoperative retention trial, the escape latency and the swimming path length of the sham-operated group were 28.4 ± 25.1 s and 562.6 ± 431.3 cm, respectively, while the escape latency and swimming path length of the MCA-occluded group were 56.7 ± 36.0 s and 1296.9 ± 812.6 cm, respectively – significantly longer those of the sham-operated rats ($p < 0.01$). MCA-occluded rats swam well and adopted a searching strategy. These was no significant difference in swimming speed between these two groups. After all the behavioral examinations were completed, rats were fixed by perfusion for neuropathological study. In five coronal sections, the area of each hemisphere of the section was determined using a digitizer and the total volume of each hemisphere was calculated. The percentage hemispheric area of the MCA-occluded side relative to the contralateral hemisphere was determined. The atrophic ratio of the MCA-occluded group was 65.1% ± 12.8%. We observed a significant correlation between the sum of three retention swimming path lengths in the Morris water maze task and the degree of brain shrinkage ($p < 0.05$).

These experiments show that focal cerebral ischemia induced by occluding the MCA causes prolonged spatial memory disturbance in rats in the chronic phase and that this correlates with shrinkage of the ischemic hemisphere.

Radial Eight-Arm Maze Task

One month after MCA occlusion, a behavioral test was carried out using a standard radial eight-arm maze task [7, 8]. After adaptation, all rats were trained once a day for 30 consecutive days, during which time a single food pellet was placed in the food cup of the eight arms. For each trial, a rat was placed on the center platform facing a randomly selected arm and was allowed to make arm choices until either all eight pellets had been eaten or 10 min had elapsed, whichever came first. The initial entry into an arm was scored as a correct choice and reentry into an arm previously visited was scored as an error. The choice accuracy was then evaluated according to two indices of maze performance: (1) the number of correct choices in the first eight choices, and (2) the number of errors during a trial. The acquisition of spatial cognition was defined as acquisition of the ability to reach at least seven different arms in the first eight choices and all eight within the first nine choices for three consecutive trials within a 30-day period. The second reacquisition session was carried out 1 month after the end of the first acquisition session. After an interval of 1 month, all rats were again tested with the radial eight-arm maze task once a day for 30 days.

When the training period for the MCA-occluded rats ($n = 30$) was started 1 month after MCA occlusion (the first session), 23 rats failed to achieve seven of eight or eight of nine correct choices for three consecutive trials during 1 month. However, in addition to the intact ($n = 6$) and sham-operated ($n = 8$) rats, the remaining seven MCA-occluded rats wee able to complete the task successfully. An analysis of variance (ANOVA) on the number of correct choices confirmed a significant difference among the three groups ($p < 0.05$). There was no significant difference in the number of correct choices between the sham-operated and intact rats. The MCA-occluded rats made more errors than either the sham-operated ($p = 0.059$) or intact rats ($p = 0.053$).

After a 1-month interval, all the rats were retrained (the second session). Nineteen MCA-occluded rats were unable to acquire spatial cognition, whereas 11 MCA-occluded rats were able to do so. In this study, the above-mentioned 19 rats were among the 23 rats which failed to perform the task successfully in the first session.

These results suggest that MCA occlusion produces long-term disturbances in spatial cognition in rats. In addition, about two-thirds of the MCA-occluded rats showed severe long-term spatial cognitive impairment. It has been proposed that normal performance in the radial eight-arm maze task requires an intact hippocampal circuitry. Damage to the hippocampus and its connections impairs spatial memory in the radial-arm maze. However, there was no evidence of any hippocampal damage in the MCA-occluded rats [8]. Therefore, it is most likely that the long-term defect in spatial cognition in these animals is due to damage of the striatum, cortex, or circuits involving these structures.

ACh Level and Impairment of Spatial Cognition

Dysfunction of the septohippocampal cholinergic system causes memory impairment in various spatial learning tasks. However, there was no evidence of any significant hippocampal damage following MCA occlusion in rats [8]. The impairment of spatial cognition following MCA occlusion may be largely due to neuronal lesions in the cerebral cortex and striatum. It has been suggested that acetylcholine (ACh) in the cortex and/or striatum is involved in the behavioral recovery of spatial cognitive impairment in rats. Therefore, we investigated the changes in cholinergic function in the chronic phase of MCA occlusion in rats, while focusing on the relationship between alterations in the levels of ACh in either the cortex or striatum and the disturbance of spatial cognition using the radial maze task.

Eleven MCA-occluded rats which were unable to acquire spatial cognition in the second session (group A), and 10 MCA-occluded rats not able to acquire spatial cognition in the second session (group B) were used for measurement of brain ACh levels. Five discrete regions, i.e., the caudate putamen (CP), parietal cortex (C4) and three regions of the frontal cortex (C1–C3), ipsilateral and contralateral to the occluded MCA were sampled and the level of ACh was measured using high pressure liquid chromatography (HPLC) with electrochemical detection [8]. Samples could not be obtained from the caudate putamen of three rats in both groups due to a deficit in this region. The ACh levels did not vary between hemispheres in the sham-operated rats. A marked decrease in the level of ACh was observed in the ipsilateral frontal cortices and CP of MCA-occluded rats. Compared with the sham-operated animals, rats in group A, which could not acquire spatial cognition in the second session, showed a decrease in the level of ACh by about 62.9% in the C2 region, 51.5% in the C3 region, and 63.6% in the CP region, while rats in group B, which could acquire spatial cognition in the second session, showed a decrease in their levels of ACh by about 63.8% in the C2 region, 49.6% in the C3 region, and 46.9% in the CP region. In group B rats, which could acquire spatial cognition in the second session, a marked increase in the level of ACh was observed in the ipsilateral and contralateral parietal (C4) regions, 163% of the mean level in the ipsilateral cortex and 225% of the level in the contralateral region in the sham operated rats ($p < 0.05$ and $p < 0.01$, respectively). The ACh level in these regions actually increased in comparison with those of either the sham-operated or intact rats. The increase in level of ACh in both the ipsilateral and contralateral C4 regions showed good correlation with the disturbance of spatial cognition ($p < 0.01$ and $p < 0.01$, respectively). However, no significant correlation was observed between the ACh level and degree of spatial cognitive deficit in the other brain regions.

It has been demonstrated that ACh plays an essential role in learning and memory [9], and the degree of depletion of ACh in the cerebral cortex and/or striatum is related to the duration of spatial cognitive deficit [4]. In this study, about two-thirds of the MCA-occluded rats showed long-term spatial cognitive impairment as judged from the level of performance in the radial eight-arm maze task. The ischemic focus, such as the ipsilateral frontal cortex and lateral caudate putamen regions, showed a marked decrease in ACh level compared with the some regions of the sham-operated rats. On the other hand, a significant increase in ACh level was observed in both the ipsilateral and contralateral parietal cortex in rats which acquired spatial cognition in the second session of this task. In addition, a close correlation was also found between the spatial cognitive deficit and ACh level in these rats. It has been reported that lesions of the nucleus basalis magnocellularis, the primary cholinergic projection to the neocortex, produces a marked disturbance in performance of the radial arm maze task [1]. This disturbance disappeared completely several weeks later, without evidence of neuronal recovery or neurochemical compensatory changes in the cholinergic system. In the case of the radial arm maze task, when a memory system is destroyed and an optimal strategy is useless to perform the task, less efficient strategies controlled by different compensatory memory systems may be used to perform the task, and thus learning will progress at a slow pace. Recently, it was reported that the impairment of spatial cognition after major cholinergic innervation of the cortex was improved by implantation of genetically modified cells that produce ACh in the parietal and frontal cortices [14]. ACh levels in the cortical areas of grafted rats with lesions of the nucleus basalis magnocellularis were significantly higher compared with the unlesioned rats. In addition, Zheng et al. [19] reported a positive turning response of the nerve growth cone in a defined extracellular gradient of ACh. From these findings and our results, we conclude that an increase in ACh level in the parietal cortex may play a role in maintaining spatial cognitive function following MCA occlusion.

Acknowledgments. This study was supported in part by a grant-in-aid for Scientific Research from the Ministry of Education, Science and Culture of Japan, and research grants from Ono Pharmaceutical Company and the Eisai Pharmaceutical Company. We are grateful to Dr. Haruyoshi Takaki (Computer Center for Research, Teikyo University) for the statistical analysis and to Mrs. Noriko Kishino and Tomomi Iwasawa for their skillful technical assistance.

References

1. Bartus RT, Flicker C, Dean RL, Pontecorvo M, Figueiredo JC, Fishcer SK (1985) Selective memory loss following nucleus basalis lesions: long term behavioral recovery despite persistent cholinergic deficiencies. Pharmacol Biochem Behav 23: 125–135
2. Fujie W, Kirino T, Tomukai N, Iwasawa T, Tamura A (1990) Progressive shrinkage of the thalamus following middle cerebral artery occlusion in rats. Stroke 21: 1485–1488
3. Hirakawa M, Tamura A, Nagashima H, Nakayama H, Sano K (1994) Disturbance of retention of memory after focal cerebral ischemia in rats. Stroke 25: 2471–2475
4. Jacobs SE, Juliano SL (1995) The impact of basal forebrain lesions on the ability of rats to perform a sensory discrimination task involving barrel cortex. J Neurosci 15: 1099–1109
5. Markgraf CG, Green EJ, Hurwitz BE, Morikawa E, Dietrich WD, McCabe PM, Ginsberg MD, Schneiderman N (1992) Sensorimotor and cognitive consequences of middle cerebral artery occlusion in rats. Brain Res 575: 238–246
6. Morris RGM, Garrud P, Rawlins JNP, O'Keefe J (1982) Place navigation impaired in rats with hippocampal lesions. Nature 297: 681–683
7. Okada M, Tamura A, Urae A, Nakagomi T, Kirino T, Mine K, Fujiwara M (1995a) Long-term spatial cognitive impairment following middle cerebral artery occlusion in rats. A behavioral study. J Cereb Blood Flow Metab 15: 505–512
8. Okada M, Nakanishi H, Tamura A, Urae A, Mine K, Yamamoto K, Fujiwara M (1995b) Long-term spatial cognitive impairment after middle cerebral artery occlusion in rats. No involvement of the hippocampus. J Cereb Blood Flow Metab 15: 1012–1021
9. Scali C, Casamenti F, Pazzagli M, Bartolini L, Pepeu G (1994) Nerve growth factor increases extracellular acetylcholine levels in the parietal cortex and hippocampus of aged rats and restores object recognition. Neurosci Lett 170: 117–120
10. Tamura A, Graham DI, McCulloch J, Teasdale GM (1981a) Focal cerebral ischemia in the rat: 1. Description of technique and early neuropathological consequences following middle cerebral artery occlusion. J Cereb Blood Flow Metab 1: 53–60
11. Tamura A, Graham DI, McCulloch J, Teasdale GM (1981b) Focal cerebral ischaemia in the rat: 2. Regional cerebral blood flow determined by [^{14}C]iodoantipyrine autoradiography following middle cerebral artery occlusion. J Cereb Blood Flow Metab 1: 61–69
12. Tamura A, Kirino T, Sano K, Takagi K, Oka H (1990) Atrophy of the ipsilateral substantia nigra following middle cerebral artery occlusion in the rat. Brain Res 510: 154–157
13. Wahl F, Allix M, Poltkine M, Boulu RG (1992) Neurological and behavioral outcomes of focal cerebral ischemia in rats. Stroke 23: 267–272
14. Winkler J, Suhr ST, Gage FH, Thall LJ, Fisher LJ (1995) Essential role of neocortical acetylcholine in spatial memory. Nature 375: 484–487
15. Yamamoto M, Tamura A, Kirino T, Shimizu M, Sano K (1988) Behavioral changes after focal cerebral ischemia by left middle cerebral artery occlusion in rats. Brain Res 452: 323–328
16. Yamamoto M, Tamura A, Kirino T, Shimizu M, Sano K (1989a) Effects of a new thyrotropin-releasing hormone derivative on behavioral changes after focal cerebral ischemia in rats. Stroke 20: 362–366
17. Yamamoto M, Tamura A, Kirino T, Hirakawa M, Shimizu M, Sano K (1989b) Effects of a new thyrotropin-releasing hormone analogue administered in rats 1 week after middle cerebral artery occlusion. Stroke 20: 1089–1091

18. Yonemori F, Yamada H, Yamaguchi T, Uemura A, Tamura A (1996) Spatial memory disturbance after focal cerebral ischemia in rats. J Cereb Blood Flow Metab 16: 973–980
19. Zheng JQ, Felder M, Connor JA, Poo M (1994) Turning of neve growth cones induced by neurotransmitters. Nature 368: 140–144

Modulation of Protein Synthesis and Calcium Uptake Following Traumatic Lesion of Rat Brain Cortex

G. MIES[1], N. KAWAHARA[2], K. KAWAI[2], G. NAGASHIMA[2], N. SAITO[2], C. RUETZLER[2], and I. KLATZO[2]

Summary

Following cold lesion of rat brain cortex, the time course of regional cerebral protein synthesis (CSP) and cerebral calcium uptake (CCU) was assessed in the same animal using [^{3}H]leucine and ^{45}CaCl by means of double tracer autoradiography. Various relationships between protein metabolism and tissue calcium homeostasis could be identified. In primary lesion sites, an enormous [^{45}Ca]CCU at variable ^{3}H-labeled tissue radioactivities indicated irreversible tissue damage. In cortex remote to the ipsilateral lesion, a transient rise in CPS was observed 2–4 days after onset of the cortical lesion without accompanying changes in CCU. In corticothalamic projection sites of the cortical lesion, however, CPS remained suppressed throughout 1–7 days post lesion with regional CCU values increasing significantly at 7 days after cold lesion of the cortex. Our results indicate that in the remote surrounding of a cortex lesion, a significant degree of cortical protein synthesis is associated with the survival and/or reorganization of neurons and/or glial cells, whereas delayed tissue damage in thalamic projection areas of the lesion site seems to be related to the persistent suppression of protein synthesis.

Introduction

Following focal injury of brain cortex, a variety of physiological, electrophysiological, and biochemical reactions occur that are associated with neuronal damage, on the one hand, or trigger the induction of protection mechanisms, on the other. After onset of cortical injury, a mismatch between blood flow and glucose metabolism has been observed for several days in the peripheral and exofocal surrounding of a lesion [12] which could contribute to the development of tissue injury. In parallel, also widespread changes in transcriptional and translational activities occur in neuronal and/or glial elements [5] which may be related to regenerative attempts of brain tissue.

Disturbance of protein synthesis and calcium homeostasis are associated with a great number of brain pathologies, e.g., focal or global cerebral ischemia and epilepsy

[1] Max-Planck-Institute for Neurological Research, Department of Experimental Neurology, Gleueler Str. 50, 50931 Cologne, Germany

[2] Stroke Branch, National Institute of Neurological Disorders and Stroke, National Institutes of Health, Building 36, Room 4D04, Bethesda, MD 20892, USA

Maturation Phenomenon in Cerebral Ischemia II
U. Ito et al. (Eds.)

[1–3, 8, 9, 11, 16, 17, 19] but little is known about alterations and interdependencies of these parameters following traumatic brain lesions. For this purpose, a double tracer autoradiographic technique was applied to determine simultaneously the cerebral calcium uptake (CCU) in vivo employing ^{45}Ca-labeled chloride as a marker of deranged cell calcium homeostasis and cerebral protein synthesis (CPS) using incorporation of [^{3}H]leucine into brain proteins as a functional measure of brain metabolism [9]. Following various periods after the onset of cold lesion of cortex, the time course of the interrelationship between local CPS and CCU was examined to identify brain regions in which predominantly regenerative processes occur and brain areas which are prone to the development of tissue injury.

Materials and Methods

Experiments were performed in male Sprague-Dawley rats (200–250 g body weight) under 1% halothane anesthesia (70% N_2O, remainder O_2). Cold injury was induced by applying a metal rod precooled in liquid nitrogen over the temporal cortex for 2 min. Animals were then allowed to survive for 1 day ($n = 3$), 2 days ($n = 3$), 4 days ($n = 3$), and 7 days ($n = 4$). Seven hours prior to the measurement of CPS, 100 μCi ^{45}Ca-labeled chloride was injected intravenously (i.v.) to assess in vivo the extent of brain tissue damage by abnormal CCU.

Animals were re-anesthetized with 1% halothane to insert catheters into the femoral vessels. Body temperature was maintained at 37°C using a feedback controlled heating pad and arterial blood gases were measured prior to the injection of tracers. For the assessment of CPS, 500 μCi [^{3}H]-4,5-leucine was injected i.v. as bolus while repeated arterial plasma samples were taken at increasing time intervals for the determination of plasma leucine specific activity. Forty-five minutes later, experiments were terminated by decapitation. Brains were immediately removed from the skull and frozen in precooled methylbutane (–50°C).

Brains were cut into 20-μm thick cryostat sections at –20°C and dried on a hot plate. Double tracer autoradiography of ^{45}Ca-labeled chloride and [^{3}H]leucine incorporated into proteins was performed as described earlier [9]. Brain sections were first subjected to ^{45}Ca-autoradiography for 2 weeks (Kodak OMC film). ^{45}Ca deposits were then removed by incubation of the sections in an aqueous solution of 0.3 *M* perchloric acid/1 m*M* ethyleneglycoltetraacetic acid (EGTA) for 2 h and defatted in 75%, 85%, and 100% alcohol for 10 min each. These pretreated sections were re-exposed to ^{3}H-sensitive film (Hyperfilm; Amersham, Braunschweig, Germany) for 2 weeks to perform ^{3}H autoradiography.

Blood samples collected during systemic [^{3}H]leucine circlulation were immediately centrifuged. Twenty microliters of plasma were reacted with 20 μl 10% trichloroacetic acid (TCA) to precipitate plasma proteins and were then centrifuged. The supernate was used to determine plasma leucine-specific activity by high pressure liquid chromatography (HPLC) analysis as described earlier [18].

Employing the in vitro principle for the quantitative measurement of a biochemical process [14], rates of cerebral protein synthesis (R_{cps}) can be calculated as described earlier [13]:

$$R_{CPS} = \frac{C^*_{PP}(T)}{\lambda \left\{ \int_0^T \left[\frac{C^*_P(t)}{C_P} \right] dt - \int_0^T \left[\frac{C^*_P(t)}{C_P} \right] e^{-(k_2^*+k_3^*+k_4^*)(T-t)} dt \right\}}$$

where $C^*_{pp}(T)$ represents ^3H brain protein radioactivity at time *T* measured autoradiographically and $C^*_p(t)/C_p$ is the plasma leucine-specific activity measured at the sampling time *t*. A value of 0.2 was taken for the sum of the constants k_2^* through k_4^*. For λ, a value equal to 0.6 was inserted [15].

Regional in vivo tissue ^{45}Ca uptake (CCU) was estimated following i.v. injection of $^{45}CaCl$ from the following ratio

$$R_{Ca} = \frac{C^*_T(T)}{C^*_P(T)}$$

where $C^*_T(T)$ represents ^{45}Ca tissue radioactivity measured autoradiographically and $C^*_P(T)$ the ^{45}Ca plasma radioactivity at time *T*. After 6h ^{45}Ca tracer circulation time, equilibrium between both compartments can be assumed [10].

Values are provided as means ± SD. Statistical comparisons were carried out by analysis of variance (ANOVA) followed by the Scheffé's test for multiple comparisons. A *p*-value < 0.05 was considered to indicate a significant difference between experimental groups.

Results and Discussion

Systemic variables of the various experimental groups remained within their physiologic range (pH, 7.3–7.5; pCO_2, 35–45 mmHg; pO_2, 120–180 mmHg).

A drastic increase in [^{45}Ca]CCU from 1.4 ± 0.2 ml/g (control) to 7.4 ± 2.1 ml/g which correlated with tissue necrosis was detected in cortex below the freezing site immediately after onset of the lesion and lasted until 7 days post lesion. At this site, ^3H tissue protein radioactivity was variable, most probably resulting from the exsudation of labeled plasma proteins following blood–brain barrier disruption caused by cortex traumatization and not from cellular amino acid incorporation into proteins.

In the close vicinity of the primary lesion, a discrete rise in [45Ca]CCU from 1.4 ± 0.2 ml/g to 2.3 + 0.6 ml/g was observed at 2 4 days post lesion without changes in cortical rates of protein synthesis from control. In more distant parts of cortex ipsilateral to the lesion, however, protein synthesis was elevated significantly between 2–4 days after the cortex lesion (control, 9.2 ± 1.7 nmol/g per minute; 2 days post lesion, 11.8 ± 1.7 nmol/g per minute, $p < 0.05$; 4 days post lesion, 12.3 ± 2.0 nmol/g per minute, $p < 0.05$) without changes in [^{45}Ca]CCU from control (Fig. 1).

In the corticothalamic projection site of cortical lesion, thalamic CPS decreased significantly from 7.7 ± 1.9 to 3.0 ± 2.2 nmol/g per minute ($p < 0.05$) at 1 day post lesion and remained suppressed throughout the whole experimental observation periods (7 days post lesion, 2.9 ± 1.9 nmol/g per minute, $p = 0.05$). During the first 2 days after the cortex lesion, no changes in tissue [^{45}Ca]CCU were observed (control, 1.4 ± 0.1 ml/g; 2

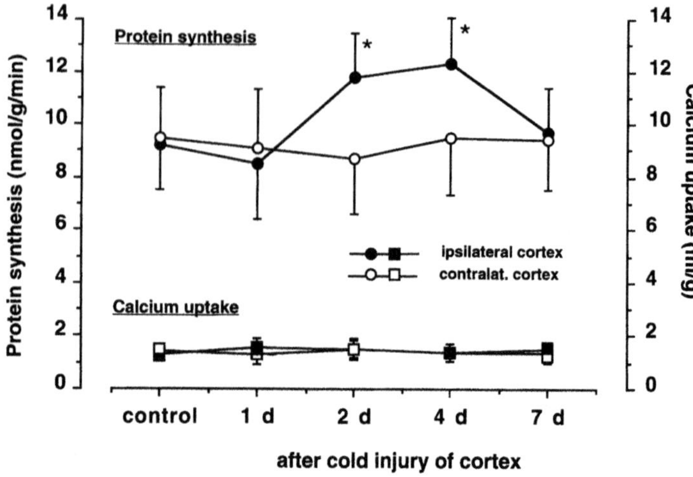

Fig. 1. During recovery from cortical cold injury, rates of protein synthesis in remote parts of ipsilateral cortex increased significantly between 2 and 4 days after cortical cold injury (*$p < 0.05$). Seven days later, however, protein synthesis rates had returned to control values. Notice that no changes in cortical calcium uptake values were observed in ipsilateral cortex remote from the cortical cold injury

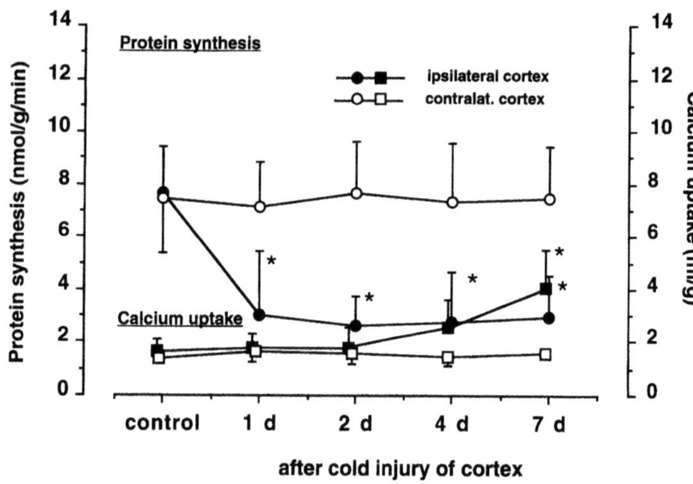

Fig. 2. After onset of cortical cold injury, rates of protein synthesis in thalamic projection areas of the ipsilateral hemisphere had decreased significantly (*$p < 0.05$) at 1 day after the lesion and remained depressed throughout the experimental survival time of animals until 7 days after cortical injury. During the first 2 days after the lesion, protein synthesis rates were suppressed without changes in tissue calcium homeostasis. A significant increase in calcium accumulation, however, occurred between 4 and 7 days after the cortex lesion

days post lesion, 1.5 ± 0.2 ml/g, n.s.) but elevated significantly at 7 days post lesion (4.1 ± 1.5 ml/g) (Fig. 2).

The results of our experimental investigation show that cerebral protein synthesis remains permanently inhibited in thalamic projection sites of the cortical lesion area. This suggests that the delayed manifestation of tissue injury in these regions as demarcated by the progressing derangement of calcium homeostasis is related to the persistent disturbance of protein metabolism as in the development of postischemic selective vulnerability [18]. In cortex remote to the primary trauma site, protein synthesis is normal or elevates at 2–4 days after focal cortical injury. Stimulation of cerebral protein synthesis seems to coincide with increased translational activity ([4, 6, 7] Kawahara et al., this volume) suggesting that a significant degree of cortical protein synthesis seems to be related to the survival and/or reorganization of neurons and/or glial cells in cortex.

References

1. Araki T, Inoue T, Kato H, Kogure K, Murakami M (1990) Neuronal damage and calcium accumulation following transient cerebral ischemia in the rat. Mol Chem Neuropathol 12: 203–213
2. Araki T, Kato H, Inoue T, Kogure K (1991) Long-term observations on calcium accumulation in postischemic gerbil brain. Acta Neurol Scand 83: 244–249
3. Dienel GA (1984) Regional accumulation of calcium in postischemic rat brain. J Neurochem 43: 913–925
4. Kawahara N, Ruetzler CA, Klatzo I (1995) Protective effects of spreading depression against neuronal damage following cardiac arrest cerebral ischaemia. Neurol Res 17: 9–16
5. Kiessling M, Gass P (1994) Stimulus-transcription coupling in focal cerebral ischemia. Brain Pathol 4: 77–83
6. Koketsu N, Berlove DJ, Moskowitz MA, Kowall NW, Caday CG, Finklestein SP (1994) Pretreatment with intraventricular basic fibroblast growth factor decreases infarct size following focal cerebral ischemia in rats. Ann Neurol 35: 451–457
7. Kraig RP, Dong L, Thisted R, Jaeger CB (1991) Spreading depression increases immunohistochemical staining of glial fibrially acidic protein. J Neurosci 11: 2187–2198
8. Mies G, Ishimaru S, Xie Y, Seo K, Hossmann K-A (1991) Ischemic thresholds of cerebral protein synthesis and energy state following middle cerebral artery occlusion in rat. J Cereb Blood Flow Metab 11: 753–761
9. Mies G, Kawai K, Saito N, Nagashima G, Nowak TS, Ruetzler CA, Klatzo I (1993) Cardiac arrest-induced complete cerebral ischemia in the rat: dynamics of postischemic in vivo calcium uptake and protein synthesis. Neurol Res 15: 253–263
10. Mulryan BJ, Neuman MW, Neuman WF, Toribara TY (1964) Equilibration between tissue calcium and injected radiocalcium in the rat. Am J Physiol 207: 947–952
11. Nagasawa H, Kogure K (1992) Alterations of 45Ca accumulation and [3H]inositol 1,4,5-trisphosphate binding using autoradiography in the exo-focal postischemic brain areas of the rat. J Neurosci Res 31: 507–512
12. Pappius HM, Wolfe LS (1983) Effects of indomethacin and ibuprofen on cerebral metabolism and blood flow in traumatized brain. J Cereb Blood Flow Metab 3: 448–459
13. Smith CB, Crane AM, Kadekaro M, Agranoff DW, Sokoloff L (1984) Stimulation of protein synthesis and glucose utilization in the hypoglossal nucleus induced by axotomy. J Neurosci 4: 2489–2496
14. Sokoloff L, Smith CB (1983) Basic principles underlying radioisotopic methods for assay of biochemical processes in vivo. In: Lambrecht RM, Rescigno A (eds) Tracer Kinetics and physiologic modeling. Springer, Berlin Heidelberg New York, pp 202–234 (Lecture notes biomathematics, vol 48)

15. Smith C, Deibler GE, Eng N, Schmidt K, Sokoloff L (1988) Measurement of local cerebral protein synthesis in vivo: influence of recycling of amino acids derived from protein degradation. Proc Natl Acad Sci USA 85: 9341–9345
16. Tomabechi M, Sako K, Yonemasu Y (1994) Accumulation of exogenous 45Ca after middle cerebral artery occlusion in rats. Neurol Med Chir 34: 76–80
17. Wasterlain CG (1974) Inhibition of cerebral protein synthesis by epileptic seizures without motor manifestations. Neurology 24: 175–180
18. Widmann R, Kuroiwa T, Bonnekoh P, Hossmann KA (1991) [14C]leucine incorporation into brain proteins in gerbils after transient ischemia: relationship to selective vulnerability of hippocampus. J Neurochem 56: 789–796
19. Yoshimine T, Hayakawa T, Kato A, Matsumoto K, Ushio Y, Mogami H (1987) Autoradiographic study of regional protein synthesis in focal cerebral ischemia with TCA wash and image subtraction techniques. J Cereb Blood Flow Metab 7: 387–393

Temperature Modulation in the Ischemic Maturation Phenomenon

F. Colbourne[1], R.N. Auer[1], and D. Corbett[2]

Summary

Brief global cerebral ischemia can result in a marked loss of hippocampal CA1 neurons which is typically delayed for a 24- to 72-h period – delayed neuronal death (DND). While CA1 necrosis is substantially reduced by intraischemic cooling, the effectiveness of postischemic hypothermia is controversial as it is thought that protection may only be transient (i.e., slow maturation). Mongolian gerbils were subjected to a severe episode (3 or 5 min) of forebrain ischemia followed, in some, by either 12 h or 24 h of hypothermia (32°C) which was initiated at 1 h following ischemia. Behavioral (memory) and histological (CA1 counts) endpoints were assessed up to 6 months postischemia. Delayed postischemic hypothermia substantially and persistently reduced both CA1 loss (especially against 3-min ischemia) and associated behavioral abnormalities. However, the 12-h cooling period was significantly less protective than the 24-h duration. In addition, some CA1 necrosis in the 12- and 24-h-treated animals occurred much later than the classical time frame expected of DND. While protracted postischemic cooling is remarkably beneficial, the amount and perhaps nature of this depends upon the duration of ischemia and cooling as well as the survival time.

Introduction

Brief global cerebral ischemia often results in selective death in several vulnerable neuronal populations such as the hippocampal CA1 cells. In humans, CA1 loss results in anterograde amnesia [26], while learning deficits in rodents are evident in tests of habituation and working memory [16, 21]. Typically, CA1 neurons undergo a delayed neuronal death (DND) which requires 24–72 h to evolve [12, 13, 18]. This slow maturation provides an opportunistic window for interventions aimed at reversing ischemic perturbations that culminate in DND.

Research over the past four decades has proven that intraischemic hypothermia is remarkably protective [3, 15, 17]. However, the efficacy of postischemic hypothermia

[1] Department of Neuroscience, Faculty of Medicine University of Calgary, 3330 Hospital Dr., N.W. Calgary, AB, Canada T2N 4N1
[2] Basic Medical Sciences, Faculty of Medicine Memorial University of Newfoundland St. John's, NF, Canada A1B 3V6

Maturation Phenomenon in Cerebral Ischemia II
U. Ito et al. (Eds.)

is highly controversial. Initial reports in humans and dogs suggested a favorable effect (e.g., reduced mortality). Cooling (30°C–34°C) was started either immediately or within a few hours and the duration ranged from a few hours to 8 days [1, 23–25]. Some of these reports, however, were tempered by the lack of appropriate controls and the absence of statistically significant benefit. Reports of deleterious effects with 29°C hypothermia (1–2 day duration) by Michenfelder et al. [14, 19, 20] further discouraged interest in hypothermia.

Recent work in rodent global ischemia models (gerbil and rat) have not entirely resolved this controversy. Busto et al. [2] found protection (3-day survival) in the rat two-vessel occlusion model (2VO, 10 min) with 3 h of immediate hypothermia (30°C), but not after a 30-min delay. Welsh and Harris [22] failed to find CA1 protection (7-day survival) with up to 2 h of immediate postischemic hypothermia (33°C and 23°C) in gerbils (5-min occlusion). Conversely, significant CA1 preservation (7-day) was found in rats (10 min of 2VO) subjected to 5 h (≈33°C) of delayed (up to 12 h) postischemic cooling [6]. Similarly, Carroll and Beek [4] reported CA1 protection (4-day) with immediate and 1-h, but not 3-h, delayed hypothermia (28°C–32°C, 2–6 h) in gerbils (5-min occlusion). Finally, Chopp et al. [5] found significant CA1 preservation (7-day) with 2 h of immediate cooling (34°C) against 8 min, but not 12 min of 2VO.

Results from Carroll and Beek [4] and Coimbra and Wieloch [6] show that neuroprotection depends upon the duration of cooling (range of 0.5–6 h) with more benefit after longer hypothermia. Thus, negative reports may have resulted from the duration of hypothermia being too short for the length of ischemia. Nonetheless, even these positive reports are challenged by the findings of Dietrich et al. [11] who reported that 3 h of immediate cooling (30°C, 10 min 2VO) only resulted in short-term CA1 rescue. Protection was found at 3- and 7-day survival, but not at 2 months. All other studies used survival times of 7 days or less. Thus, from these studies it is not clear if postischemic hypothermia can protect chronically.

The aim of the present investigation [8, 9] was three-fold. Firstly, is more protracted hypothermia (12 h or 24 h, 32°C) persistently beneficial (10- to 180-day survival)? Secondly, does this protection depend upon the duration of ischemia (3 vs 5 min in gerbil)? Finally, does postischemic hypothermia provide functional (habituation) as well as histological (CA1) benefit?

Materials and Methods

Subjects. Female Mongolian gerbils (≈55 g) were housed individually with free access to food (rat chow) and water throughout the experiment. All procedures were in accordance with the Canadian Council on Animal Care guidelines. In Experiment 1 [8], groups were HYPO-5-10, HYPO-5-30, HYPO-3-10, HYPO-3-30, ISCH-5-10, ISCH-5-30, ISCH-3-10 and ISCH-3-30 ($n = 7$–9 per group). The numbers refer to the duration of ischemia (3 or 5 min) and the survival time (10 or 30 days), respectively. In Experiment 2 [8, 9], groups were SHAM-30 ($n = 4$), ISCH-30 ($n = 8$), HYPO-30 ($n = 8$), SHAM-180 ($n = 10$), ISCH-180 ($n = 14$) and HYPO-180 ($n = 9$). Ischemic groups were subjected to 5 min of ischemia. The numbers refer to the survival time (days).

Brain Temperature. Brain temperature (dorsal striatum) was measured by wireless AM probes (XM-FH, Mini-Mitter Co., Inc., Sunriver, OR, USA) that were inserted into a guide cannula. This was a 5.0-mm tube that was implanted 4 days prior to ischemia/sham operation. Brain probes were first inserted (3 h of baseline temperature) 2 days prior to ischemia.

Ischemia. Gerbils were anesthetized with 1.5% halothane (70% N_2O, 30% O_2) followed by insertion of the brain probe. Animals were then wrapped in a homeothermic body blanket and a heated water blanket which enveloped the head. Following a ventral neck incision, the carotid arteries were isolated and occluded (except SHAMs) for either 3 or 5 min. After verification of reflow, animals were sutured and returned to cages for continuous brain temperature measurement (every 20 s).

Hypothermia. Animals were shaved (back and abdomen) at 1 h postischemia under brief halothane anesthesia. Cooling was manually induced and maintained (32.0°C ± ≈0.2°C) in awake, freely-moving gerbils by the frequent use of an overhead fan and cold water spray (≈4°C). Heat was added by an overhead 100-watt lamp. Gerbils were cooled for a total of 12 h (Experiment 1) or 24 h (Experiment 2). The rate of cooling/rewarming was 1.0°C per 10 min.

Open Field Testing. In Experiment 2, gerbils were exposed to an open field (72 cm × 76 cm × 57 cm) on days 5, 10, 30 and 180, depending upon survival time. The number of squares crossed (25 in box) per session reflects habituation to novelty and is sensitive to hippocampal CA1 injury [21].

Histology. Gerbils were sacrificed with an overdose of Somnotol and fixed with 10% formalin. In experiment 1, 10-μm frozen sections were collected and stained with cresyl violet. The number of viable CA1 neurons were counted and summated in medial, middle and lateral sectors (−1.7 mm to bregma) and compared to normal animals from Colbourne and Corbett [7]. In Experiment 2, paraffin embedded brains were cut at 6 μm and stained with haematoxylin and eosin. Again, CA1 cells were counted in medial, middle and lateral sectors, and an analysis of variance (ANOVA, with specific comparisons) of sector differences was made.

Results

All groups had similar baseline brain temperatures (≈36.3°C) and their intraischemic temperatures were kept close to normothermia. Hypothermic and ischemic control groups were also similar in the first postischemic hour.

Experiment 1. Untreated ischemia of 3-min (≈14% of SHAM) and especially 5-min durations (≈1%) resulted in near-total CA1 loss with 10- and 30-day survival. The 12-h hypothermic period significantly ($p < 0.01$) reduced ischemic CA1 loss in all comparisons (e.g., HYPO-5-10 vs CONT-5-10). CA1 protection against 3 min of ischemia was almost identical at 10 days (≈79% of SHAM) and 30 days (≈85%), indicating

Table 1. CA1 data (% normal ± SD in medial, middle and lateral sectors) for hypothermic groups in Experiment 2

Group	Medial	Middle	Lateral
HYPO-30	87.21 ± 14.29	91.27 ± 3.65	93.57 ± 7.33
HYPO-180	51.73 ± 26.73	73.86 ± 37.07	85.13 ± 37.39

persistent benefit ($p = 0.52$). Protection was also markedly better against 3 min than 5 min of ischemia where protection appeared to decline from 10 days (≈35% of SHAM) to 30 days (≈12%) ($p = 0.053$).

Experiment 2. Untreated ischemic gerbils had higher activity levels (impaired habituation) in the open field on each test day compared to both SHAM ($p < 0.01$) and HYPO groups ($p < 0.02$). Nonetheless, HYPO gerbils were significantly impaired (vs SHAM) on days 5 ($p < 0.04$) and 10 ($p < 0.01$).

Untreated ischemia (5-min) consistently resulted in near total CA1 loss (≈2% of SHAM in all sectors) at 30- and 180-day survival. Hypothermia (24-h) significantly ($p < 0.01$) and almost totally reduced CA1 loss against 5 min of ischemia at 30 days (Table 1). This was similar in all sectors. Protection at 6 months, however, was regional with greater benefit in more lateral sectors (e.g., medial vs lateral, $p < 0.01$). Importantly, protection in medial, but not middle or lateral, CA1 was significantly less at 6 months than at a 1-month survival ($p < 0.01$), illustrating a very slow DND in medial CA1.

Discussion

We conclude that delayed postischemic hypothermia is persistently, and perhaps permanently, neuroprotective in CA1 with an associated attenuation of habituation impairments. It is clear, however, that protection was influenced by the duration of ischemia and the length of cooling. Understandably, hypothermia was much more efficacious against 3 vs 5 min of ischemia (see also [5]). More importantly, these results illustrate the value of prolonged cooling as a 12-h hypothermic period was much less effective than a 24-h duration against 5 min of ischemia. Furthermore, delayed cooling for less than a 6-h period is ineffective (unpublished data).

Maturation of CA1 DND (especially medial CA1) after postischemic hypothermia is not fixed, in agreement with Dietrich et al. [11]. Our results support their conclusion that brief hypothermia (depending upon severity of ischemia) may not be chronically beneficial. However, the robust functional and histological benefit presently obtained against 3 min of ischemia with 12-h cooling and against a 5-min occlusion with 24-h hypothermia indicates that postischemic hypothermia can be a true neuroprotectant like intraischemic cooling.

Other data [4, 6, 9] also show that hypothermic protection depends upon the intervention delay and the degree of cooling. Thus, further work is needed to systematically determine the optimal degree and duration of hypothermia required to offset global and focal ischemic injury (in progress).

We conclude that postischemic temperature must be controlled for in ischemia research such as with an automated brain temperature control system [10]. Secondly, putative neuroprotectants should be assessed with both behavioral and long-term histological endpoints. Finally, protracted mild hypothermia deserves clinical investigation.

Acknowledgments. The authors acknowledge research support by the Medical Research Council of Canada awarded to D. Corbett. F. Colbourne was supported by a MRC studentship.

References

1. Benson DW, Williams GR, Spencer FC, Yates AJ (1959) The use of hypothermia after cardiac arrest. Anesth Analg 38: 423–428
2. Busto R, Dietrich WD, Globus MY-T, Ginsberg MD (1989) Postischemic moderate hypothermia inhibits CA1 hippocampal ischemic neuronal injury. Neurosci Lett 101: 299–304
3. Busto R, Dietrich WD, Globus MY-T, Valdés I, Scheinberg P, Ginsberg MD (1987) Small differences in intraischemic brain temperature critically determine the extent of ischemic neuronal injury. J Cereb Blood Flow Metab 7: 729–738
4. Carroll M, Beek O (1992) Protection against hippocampal CA1 cell loss by post-ischemic hypothermia is dependent on delay of initiation and duration. Metab Brain Dis 7: 45–50
5. Chopp M, Chen H, Dereski MO, Garcia JH (1991) Mild hypothermic intervention after graded ischemic stress in rats. Stroke 22: 37–43
6. Coimbra C, Wieloch T (1994) Moderate hypothermia mitigates neuronal damage in the rat brain when initiated several hours following transient cerebral ischemia. Acta Neuropathol (Berl) 87: 325–331
7. Colbourne F, Corbett D (1992) Effects of d-Amphetamine on the recovery of function following cerebral ischemic injury. Pharmacol Biochem Behav 42: 705–710
8. Colbourne F, Corbett D (1994) Delayed and prolonged post-ischemic hypothermia is neuroprotective in the gerbil. Brain Res 654: 265–272
9. Colbourne F, Corbett D (1995) Delayed postischemic hypothermia: a six month survival study using behavioral and histological assessments of neuroprotection. J Neurosci 15: 7250–7260
10. Colbourne F, Sutherland GR, Auer RN (1996) An automated system for regulating brain temperature in awake and freely-moving rodents. J Neurosci Methods 67: 185–190
11. Dietrich WD, Busto R, Alonso O, Globus MY-T, Ginsberg MD (1993) Intraischemic but not postischemic brain hypothermia protects chronically following global forebrain ischemia in rats. J Cereb Blood Flow Metab 13: 541–549
12. Ito U, Spatz M, Walker JT, Klatzo I (1975) Experimental cerebral ischemia in Mongolian gerbils. Acta Neuropathol (Berl) 32: 209–223
13. Kirino T (1982) Delayed neuronal death in the gerbil hippocampus following ischemia. Brain Res 239: 57–69
14. Michenfelder JD, Milde JH (1977) Failure of prolonged hypocapnia, hypothermia, or hypertension to favorably alter acute stroke in primates. Stroke 8: 87–91
15. Nurse S, Corbett D (1994) Direct measurement of brain temperature during and after intraischemic hypothermia: correlation with behavioural, physiological and histological endpoints. J Neurosci 14: 7726–7734
16. Ordy JM, Thomas GJ, Volpe BT, Dunlap WP, Colombo PM (1988) An animal model of human-type memory loss based on aging, lesion, forebrain ischemia, and drug studies with the rat. Neurobiol Aging 9: 667–683
17. Pontius RG, Bloodwell RD, Cooley DA, De Bakey ME (1954) The use of hypothermia in the prevention of brain damage following temporary arrest of cerebral circulation: experimental observations. Surg Forum 5: 224–228
18. Pulsinelli WA, Brierley JB, Plum F (1982) Temporal profile of neuronal damage in a model of transient forebrain ischemia. Ann Neurol 11: 491–498

19. Steen PA, Milde JH, Michenfelder JD (1980) The detrimental effects of prolonged hypothermia and rewarming in the dog. Anesthesiology 52: 224–230
20. Steen PA, Soule EH, Michenfelder JD (1979) Detrimental effect of prolonged hypothermia in cats and monkeys with and without regional cerebral ischemia. Stroke 10: 522–529
21. Wang D, Corbett, D (1990) Cerebral ischemia, locomotor activity and spatial mapping. Brain Res 533: 78–82
22. Welsh FA, Harris VA (1991) Postischemic hypothermia fails to reduce ischemic injury in gerbil hippocampus. J Cereb Blood Flow Metab 11: 617–620
23. Williams GR, Spencer FC (1958) The clinical use of hypothermia following cardiac arrest. Ann Surg 148: 462–466
24. Wolfe KB (1960) Effect of hypothermia on cerebral damage resulting from cardiac arrest. Am J Cardiol 6: 809–812
25. Zimmerman JM, Spencer FC (1959) The influence of hypothermia on cerebral injury resulting from circulatory occlusion. Surg Forum 9: 216–218
26. Zola-Morgan S, Squire LR, Amaral DG (1986) Human amnesia and the medial temporal region: enduring memory impairment following a bilateral lesion limited to field CA1 of the hippocampus. J Neurosci 6: 2950–2967

The Role of Endothelin-1 and Nitric Oxide in Cerebral Ischemia

M. Spatz, Y. Yasuma, A. Strasser, and R.M. McCarron

Summary

The effect of cerebral ischemia/reperfusion on vascular responses and endothelin (ET) content in cerebrospinal fluid was investigated in the presence and absence of Nω-nitro-L-arginine (NLA), the nitric oxide (NO) inhibitor. A single intraperitoneal injection of NLA (4h prior to the induction of ischemia) stereospecifically increased systemic blood pressure (SBP) and reduced the transient, initial (early) recovery of cerebral blood flow (CBF). However, NLA treatment did not significantly affect the late hypoperfusion observed in either Ringer's- or Nω-nitro-D-arginine methyl ester (D-NAME)-treated animals. Both the early and late hypoperfusion coincided with the increased content of ET-1 in the cerebrospinal fluid. These findings strongly suggest that endogenously released ET-1 is responsible for the observed postischemic hypoperfusion.

Introduction

In the last decade, the discovery of endothelin (ETs) and nitric oxide (NO) provided a new field of research that is important for understanding the function of many organs under physiological and pathological conditions [1, 4, 6, 9, 10, 15, 18]. Initially, the peripheral vascular endothelium was identified as the source of ET and NO secretion [9, 15, 23]. However, the synthesis of ETs and NO has also been demonstrated in a variety of other cellular elements of many organs including brain [5, 13, 14, 18].

Three separate human ET genes were recognized encoding three distinct ET peptides (ET-1, ET-2 and ET-3). Each member of the ET family contains 21 amino acid residues, including four cysteine residues forming two intermolecular disulfide bonds. ET-1 is formed enzymatically from pre-pro-ET-1 and pro-ET-1 [18, 23].

NO is synthesized through oxidation of L-arginine by a family of nitric oxide synthases (NOS). At least three separate products of NOS have been identified. All these isoforms of NOS are flavoproteins which require nicotinamide adenine dinucleotide, reduced (NADPH) and tetrahydrobiopterin as cofactors for their activity. They are either constitutively expressed and calcium-dependent (cNOS) or inducible and calcium-independent (iNOS)[15].

Stroke Branch, National Institute of Neurological Disorders and Stroke, National Institutes of Health, 36 Convent Drive, MSC 4128, Bethesda, MD 20892, USA

Maturation Phenomenon in Cerebral Ischemia II
U. Ito et al. (Eds.)

It is generally accepted that both ET-1 and NO (among a number of other substances produced by vascular endothelium) play pivotal roles in the autocrine-paracrine regulation of vascular and microvascular function in brain and other organs. NO represents one of the various endothelial vasoactive agents or factors (e.g., prostacyclin and adenosine) that are considered to be cytoprotective. In contrast, ET-1 belongs to another group of endothelial substances which, when formed and released in excess may impair perfusion, alter the blood–brain barrier (BBB) permeability and/or mediate cellular injury (e.g., angiotensin II, thromboxane, leukotrienes, platelet-activating factor and superoxide radicals). In addition, there is ample evidence of existing "cross-talk" among several endothelial-formed substances which can be modulated by blood-borne or brain-derived agents (e.g., hormones, neurotransmitters, and peptides). Thus, under- or over-production of one of the endothelial mediators may primarily or secondarily contribute to the development of cerebrovascular dysfunction or diseases [12, 16]. In particular, many studies have suggested an existing delicate balance between the endothelial content of ET-1 (most potent known vasoconstrictor) and NO (the powerful vasodilator) which play an important role in maintaining vascular tone, blood pressure and cerebral blood flow (CBF). Furthermore, changes in either the content or responses to ET-1 and/or NO have been described in cerebral ischemia, but none have dealt with their possible interaction in ischemic brain until recently.

This report describes our latest data suggestive of a functional relationship between ET-1 and NO in experimental ischemia/reperfusion.

Materials and Methods

Mongolian gerbils (3-month-old females, weighing 50–70 g) obtained from Tumble-Brook Farm (West Brookfield, MA) were used for these studies. Animals were housed under 12-h light–dark cycle with free access to food and water for 4–7 days before the experiments. As reported previously, separate groups of animals (five to 12 animals in each) under halothane anesthesia (1.5%) and N_2O/O_2 (2:1) and spontaneous ventilation were subjected to bilateral carotid artery occlusion (15 min) alone or with release (5, 15, and 120 min)[20, 21]. Respective groups of sham-operated animals served as controls. Systemic blood pressure (SBP), using Universal Harvard Oscillograph (Kent, UK) and blood pressure transducer, CBF, head (temporal muscle) and rectal temperature were continuously monitored and maintained at 37°C–38°C with a thermostatically regulated heating lamp (Model 73, Yellow Spring Instruments Co., Inc., Yellow Spring, OH, USA) during the entire experimental procedure. The severity of the ischemic insult and the pattern of the CBF were assessed by measuring CBF using laser Doppler flowmetry (Laserflo Model BPM 403A, TSI; St. Paul, MN, USA). The treatment consisted of intraperitoneal (i.p.) administration of Nω-nitro-L-arginine (NLA) or Nω-nitro-D-arginine methyl ester (D-NAME) (40 mg/kg dissolved in 0.5 ml of Ringer's solution) 4 h prior to induction of ischemia or sham operation [20, 21]. Cerebrospinal fluid (CSF) was aspirated from the cisterna magna under a dissecting microscope at the end of each designated experimental period. The samples of CSF were immediately centrifuged and 30-ml aliquots were frozen at −70°C until assayed. The level of ET-1-like immunoreactivity in the CSF was measured using a commercially available kit (Peninsula Laboratories, Belmont, CA). The nonspecific binding of tracer was 4% of the radiolabel added; the sensitivity of the radioimmunoassay (RIA) was 1–128 pg/tube. All procedures were performed strictly in accordance with the NIH *Guide for Care and Use of Laboratory Animals* and approved by the NIH Animal Care and Use Committee.

Statistical Analysis. All obtained data are presented as means ± SEM. Statistical differences were evaluated with analysis of variance (ANOVA) alone or followed by Fisher's protected least square difference (PLSD) test between particular experimental groups. A value of $p < 0.01$ was considered statistically significant.

Results

General Effects of Ischemia/Reperfusion in Ringer's or D-NAME-Treated Gerbils

Temperature. Neither ischemia nor ischemia–reperfusion significantly affected the temperature (rectal and temporal muscle). A similar slight decrease in the temperature of the temporal muscle was seen in both groups [e.g., Ringer's or D-NAME 35.5°C ± 1.1°C and 32.5°C ± 1.1°C, mean ± SD ($n \pm 6$) in preischemic, ischemia/reperfusion, respectively).

Systemic Blood Pressure. An incease (41%–59%) in SBP above control level was seen during ischemia only in all gerbils irrespective of treatment.

Cerebral Blood Flow. Ischemia caused a similar reduction in CBF to <5% of the preischemic level (Fig. 1). An almost complete recovery of the CBF was observed at 7.5 min of reperfusion. Subsequently, the CBF progressively declined (7.5–15 min) resulting in prolonged hypoperfusion.

Effect of Ischemia/Reperfusion on CSF ET-1 Content in Ringer's or D-NAME-Treated Gerbils

Ischemia alone had no effect on the content of ET-1 in the CSF (Fig. 2A). However, the level of ET-1 doubled and tripled after 5 min and 15 min of reperfusion, respectively. Subsequently, normal values of CSF ET-1 content were found after 2 h of reperfusion.

General Effects of Ischemia/Reperfusion in NLA-Treated Gerbils

Temperature. There was no significant difference in the rectal or temporal muscle temperature seen between NLA-treated and Ringer's- or D-NAME-treated animals.

Systemic Blood Pressure. NLA significantly raised the SBP above the Ringer's- and D-NAME-treated gerbils during pre- and postischemia. The SBP level varied between

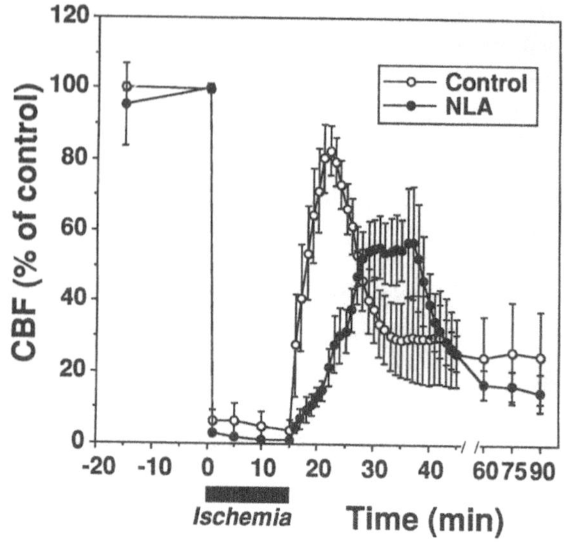

Fig. 1. Effect of Nω-nitro-L-arginine (NLA) on the cerebral blood flow (CBF) during ischemia-reperfusion in Mongolian gerbils. Note: the postischemic hypoperfusion is shown only up to 75 min but the same levels were seen at 120 min

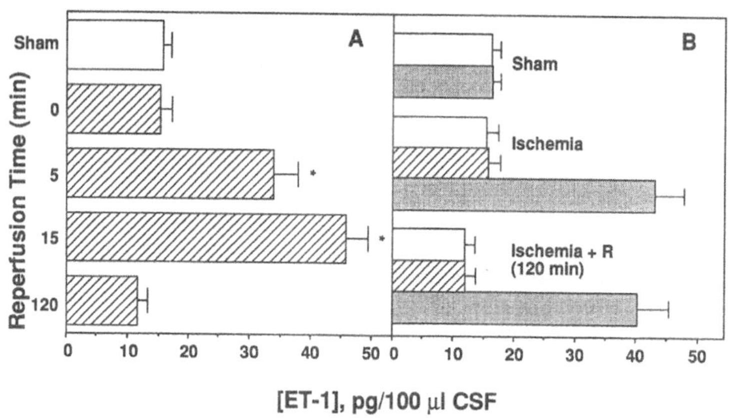

Fig. 2. A The effect of bilateral cerebral ischemia-reperfusion (15 min, 120 min respectively) on the content of endothelin-1 (ET-1) in the cerebrospinal fluid (CSF) in Mongolian gerbils. Effect of intraperitoneally administered physiological saline on sham-operated animals (□) and on gerbils subjected to ischemia/reperfusion (□). Each *bar* represents mean ± SEM for five to 12 animals. *Significant difference ($p < 0.01$; one-way analysis of variance, ANOVA) as compared to sham-operated animals. **B** The effects Nω-nitro-L-arginine (NLA) on the content of endothelin-1 (ET-1) in the cerebrospinal fluid (CSF) in Mongolian gerbils during bilateral cerebral ischemia/reperfusion (15 min, 120 min respectively). Effect of intraperitoneally administered physiological saline (□) or 40 mg/kg b.w. of Nω-nitro-L-arginine (▦) on the levels of ET-1 in CSF during brain ischemia and ischemia/reperfusion or in sham-operated animals (□). Each *bar* represents mean ± SEM for six to 12 animals. *Significant difference ($p < 0.01$; ANOVA) as compared to the corresponding group of sham-operated animals

76.07 ± 1.88 and 85.83 ± 1.75 mm Hg in NLA-treated animals as compared to variable levels of 66.7 ± 3.5 and 75.8 ± 1.77 mm Hg in Ringer's- or D-NAME-treated animals. However, the NLA had no additional effect on the ischemic elevation of SBP beyond that seen in Ringer's- or D-NAME-treated gerbils.

Cerebral Blood Flow. The ischemic reduction in CSF with NLA was not different from that observed in Ringer's or D-NAME treatment (Fig. 1). However, the initial transient recovery seen in Ringer's- or D-NAME-treated gerbils was not seen in NLA-treated animals. At 15 min of reperfusion there was no significant difference between the CBF with NLA and either Ringer's or D-NAME treatment.

Effect of Ischemia/Reperfusion on CSF ET-1 Content in NLA-Treated Gerbils

The ET-1 content in the CSF was markedly increased after ischemia alone and remained high (about three-fold above sham or Ringer's-treated gerbils) at 2-h reperfusion (Fig. 2B).

Discussion

These findings indicate that NLA, the inhibitor of NOS, affects the initial recovery of CBF during reperfusion after global ischemia in gerbils. The changes in CBF coincided with an increased ET-1 content in CBF. The observed increase of the pre- and postischemic level of SBP and the reduced recovery of CBF induced by ischemia–reperfusion in animals pretreated with NOS inhibitors was previously described by us and other investigators in global and focal ischemia [17, 20, 21].

NLA, among other L-enantiomers, has been widely used as an NOS inhibitor of both constitutive (endothelial and neuronal NOS isoforms) and inducible enzyme activities. These substances, but not the D-enantiomers, block the conversion of arginine to citrulline without affecting the cGMP response induced by NO donors, (e.g., nitroprusside) [15]. Thus, the ineffectiveness of D-NAME in contrast to NLA to alter the pre- and postischemic SBP and, most importantly, the early, transient postischemic recovery of the CBF indicate involvement of NO in the NLA-induced elevation of SBP and decreased CBF recovery.

The reported increased ET-1 content in plasma, CSF or tissue in patients and/or animals with subarachnoid hemorrhage, vasospasm hypertension and brain infarct implicated ET-1 in the pathogenesis of cerebrovascular diseases [2, 11, 18, 22]. This supposition was strengthened by induction of brain infarct with exogenous ET-1. The results reported here confirm for the first time the assumption that cerebral ischemia raises the ET-1 content in CSF [14].

As we mentioned above, NO production has been considered as neuroprotective and neurotoxic. Both NO properties have been implicated as playing a role in brain ischemia depending on the type, time and concentration of administered NOS inhibitors and the model of ischemia [10]. However, studies in mutant mice deficient in neuronal synthase activity clarified the role on NO in focal cerebral ischemia [8]. Middle cerebral artery (MCA) occlusion (24 h and 72 h) induced a smaller infarct and a lesser neurological deficit in mutant mice as compared to normal mice. In addition, the lesion size in the same mutant mice became larger after inhibition of endothelial NOS by NLA. Based on these observations, it appears that neuronal NO exacerbates, whereas vascular NO protects, ischemic brain injury following MCA occlusion.

It has been suggested that NO, in addition to its involvement as an amplifier in glutamate-mediated tissue injury, may react with superoxide anions (O_2^-) to form peroxynitrite ($ONOO^-$)[4]. The decomposition of these compounds yields hydroxyl free radicals and nitrogen dioxide which are implicated in lipid peroxidation and neurotoxicity.

Taking into account all these observation, it is clear that either ET-1 or NO are involved in mechanisms of brain injury. As far as ET-1 is concerned, its contribution to tissue injury may be secondary due to its effect on SBP and CBF. Based on in vitro studies, it is also possible that other vascular responses to ET-1, namely stimulation of brain capillary permeability and ionic transport systems, as well as increased adhesive interactions between endothelium and leukocytes, could participate in the events of cerebrovascular ischemia [19]. Most of these results were obtained by investigating the effect of exogenous ET-1. Similarly, the investigations of interaction between ET-1 and NO were conducted with the use of exogenous ET-1. Donors of NO were reported to inhibit ET-1 secretion [3, 12], whereas exogenous ET-1 was thought to stimulate endothelial NO production [7]. Since both the ET-1 and NO synthetic-metabolic pathways are also present in cells other than endothelium, the prolonged elevation of ET-1 in CSF induced by NLA in ischemia/reperfusion could be due to vascular and/or neuronal changes. However, the exact cellular origin and mechanism responsible for both the transient increase of ET-1 in CSF and the NLA amplification of ET-1 content in CSF in ischemia/reperfusion are unknown. Based on these findings and previous observations that also demonstrated an association of the incomplete recovery of postischemic CSF, elevation of SBP and changes in dopamine metabolism in the same model of ischemia [21], we suggest that the changes are more likely resulting primarily from vascular rather than neuronal NLA effects. We also suggest that the endogenously released ET-1 is responsible for the observed hypoperfusion in ischemia/reperfusion irrespective of treatment. This hypothesis has lately been tested in the same ischemic model using an ETA receptor antagonist which blocked many functional biochemical properties of cultured endothelium derived from human and rat brain capillaries. These investigations demonstrated that NLA reduction of initial, transient CBF recovery and the late postischemic hypoperfusion are fully reversed by treatment with BQ123, the known specific ETA receptor antagonist for which ET-1 is the endogenous ligand. In conclusion, the findings of this report provide additional support implicating ET-1 and NO involvement in brain ischemia.

References

1. Bacic F, Uematsu S, McCarron RM, Spatz M (1992) Secretion of immunoreactive endothelin-1 by capillary and microvascular endothelium of human brain. Neurochem Res 17: 699–702
2. Barone FC, Globus MY, Price WJ, White RF, Storer BL, Feuerstein GZ, Busto R, Ohlstein EH (1994) Endothelin levels increase in rat focal and global ischemia. J Cereb Blood Flow Metab 14: 337–342
3. Boulanger C, Luscher T (1990) Release of endothelin from the porcine aorta. Inhibition by endothelin-derived nitric oxide. J Clin Invest 85: 587–590

4. Dawson DA (1994) Nitric oxide and focal cerebral ischemia: multiplicity of actions and diverse outcome. Cerebrovasc Brain Metab Rev 6: 299–324
5. Ehrenreich H, Lange M, Near KA, Anneser F, Schoeller LA, Schmid R, Winkler PA, Kehrl JH, Schmiedek P, Goebel FD (1992) Long term monitoring of immunoreactive endothelin-1 and endothelin-3 in ventricular cerebrospinal fluid, plasma, and 25-h urine of patients with subarachnoid hemorrhage. Res Exp Med 192: 257–268
6. Haynes WG, Davenport AP, Webb DJ (1993) Endothelin: progress in pharmacology and physiology. Trends Pharmacol Sci 14: 225–228
7. Hirata Y, Emori T (1993) Cellular mechanism of endothelin-induced nitric oxide synthesis by cultured bovine endothelial cells. J Cardiovasc Pharmacol 8: S225–S228
8. Huang Z, Huang PL, Panahian N, Dalkara T, Fishman MC, Moskowitz MA (1994) Effects of cerebral ischemia in mice deficient in neuronal nitric oxide synthase. Science 265: 1883–1885
9. Hunley TE, Iwasaki S, Homma T, Kon V (1995) Nitric oxide and endothelin in pathophysiological settings. Pediatr Nephrol 9: 235–424
10. Iadecola C, Pelligrino DA, Moskowitz MA, Lassen NA (1994) Nitric oxide synthase inhibition and cerebrovascular regulation. J Cereb Blood Flow Metab 14: 175–192
11. Khraibi AA, Heublein DM, Knox FG (1993) Increased plasma level of endothelin-1 in the Okamoto spontaneously hypertensive rat. Mayo Clin Proc 68: 42–46
12. Luscher TF (1992) Heterogeneity of endothelial dysfunction in hypertension. Eur Heart J 13 [Suppl D]: 50–55
13. MacCumber MW, Ross AC, Snyder SH (1990) Endothelin in the brain: receptors, mitogenesis and biosynthesis in glial cells. Proc Natl Sci USA 87: 2359–2363
14. Macrae IM, Robinson MJ, Graham DI, Reid JL, McCulloch J (1993) Endothelin-1 induced reductions in cerebral blood flow: dose dependency, time course, and neuropathological consequences. J Cereb Blood Flow Metab 13: 276–284
15. Moncada S, Palmer RM, Higgs EA (1991) Nitric oxide: physiology, pathophysiology, and pharmacology. Pharmacol Rev 43: 109–142
16. Panza JA, Quyyumi AA, Brush JE Jr, Epstein SE (1990) Abnormal endothelin-dependent vascular relaxation in patients with essential hypertension. N Engl J Med 323: 22–27
17. Prado R, Watson BD, Webster P (1993) Effects of nitric oxide synthase inhibition on cerebral blood flow following bilateral carotid artery occlusion and recirculation in the rat. J Cereb Blood Flow Metab 13: 720–723
18. Rubanyi GM, Polokoff MA (1994) Endothelins: molecular biology, biochemistry, pharmacology, physiology, and pathophysiology. Pharmacol Rev 46: 325–415
19. Spatz M, Stanimirovic D, Strasser A, McCarron RM (1995) Nitro-L-arginine augments the endothelin-1 content of cerebrospinal fluid induced by cerebral ischemia. Brain Res 684: 99–102
20. Spatz M, Stanimirovic D, Strasser A, McCarron RM (1995) Endothelin as a mediator of blood-brain barrier function. In: Greenwood J (ed) New concepts of the blood-brain barrier. Polonium, Amsterdam, pp 47–61
21. Strasser A, Yasuma Y, McCarron RM, Ishii H, Stanimirovic D, Spatz M (1994) Effect of nitro-L-arginine on cerebral blood flow and monoamine metabolism during ischemia/reperfusion in the Mongolian gerbil. Brain Res 664: 197–201
22. Suzuki R, Masaoka H, Hirata Y, Marumo F, Isotani E, Hirakawa K (1992) The role of endothelin-1 in the origin of cerebral vasospasm in patients with aneurysmal subarachnoid hemorrhage. J Neurosurg 77: 96–100
23. Yanagisawa M, Inoue A, Takuwa Y, Mitsui Y, Kobayashi M, Masaki T (1989) The human preproendothelin-1 gene: possible regulation by endothelial phosphoinositide turnover signaling. J Cardiovasc Pharmacol 13 [Suppl 5]: S13–S17

Microglial Cell Death Following Phagocytosis of Zymosan-A Under a Video-Enhanced Contrast – Differential Interference Contrast Microscope: Does This Include Apoptosis?

M. Tomita, Y. Fukuuchi, N. Tanahashi, M. Kobari, H. Takeda, S. Konno, and M. Yokoyama

Summary

Employing a novel video-enhanced contrast-differential interference contrast (VEC-DIC) microscope with a mercury light source, the process of microglial cell death after phagocytosis of zymosan-A was examined for its morphological changes, especially of the nucleus. The materials used were microglia (exp. $n = 15$) isolated from the new-born rat brain. Zymosan-A was introduced into the chamber on the stage of the microscope where the microglia were being observed. Aggregates of zymosan-A became partially or totally engulfed by the microglia. At ca. 10–30 min of phagocytosis, the microglia began to swell, with ballooning of the cell membrane, and then burst. The cell body changes in terms of the circumferential length were: 84.38 μm in the control and 141.71 μm at the time of bursting ($p < 0.05$). Those in the area of the cell body were: 36.43 μm^2 in the control and 47.92 μm^2 at the time of bursting ($p < 0.05$). The average dimensional changes of the nucleus ($n = 5$, where the nuclear changes could be followed) in the control, at the time before bursting, after bursting, and at 10 min after bursting were 7.45, 7.32, 7.25, respectively, and 7.01 μm for the long axis, and 6.53, 6.13, 6.81, respectively, and 6.55 μm for the short axis. The minimal change in nuclear size together with swelling/dissolution of the plasma membrane suggests that the self-inflicted cell death triggered by zymosan-A, and possibly facilitated by the UV light, might result from necrosis due to self-inflicted effects of autogenous toxic substances. However, the findings were not conclusive since chromatin condensation with cell shrinkage was observed in some cases, suggestive of apoptosis.

Introduction

The maturation of cerebral ischemia and recovery may be greatly modified by microglia. The active forms of microglia–macrophages, on the one hand, contribute to the repair process of the ischemic brain as scavengers by engulfing debris and by secreting astroglial growth factors, and, on the other hand, aggravate the degenerative process by releasing cytotoxins which kill neighboring neuronal cells [1]. The microglia–macrophages usually die following completion of phagocytosis. The question arises as to whether or not such microglial cell death after fulfillment of the

Department of Neurology, School of Medicine, Keio University, 35 Shinanomachi, Shinjuku-ku, Tokyo 160, Japan

Maturation Phenomenon in Cerebral Ischemia II
U. Ito et al. (Eds.)

scavenging duty is due to apoptosis. In this study, employing a novel video enhanced contrast-differential interference contrast (VEC-DIC) microscope, we examined the morphological changes of the nucleus and cell body of microglia which accompanied cell death after phagocytosis of zymosan-A, a polysaccharide derived from the cell walls of yeast.

Materials and Methods

We used the VEC-DIC microscope in a modification of Terakawa's original setup [5], involving an assembly of three systems: an optical system, an electrical system and a video system [6]. The optical system consisted of an inverted Nomarski microscope (Axiovert 135; Carl Zeiss, Jena, Germany) equipped with a ×100 DIC objective lens and two insertion lenses of ×2.5 and ×4 magnification. Optical images were obtained employing a charge-coupled device (CCD) camera (3077, Hamamatsu Photonics; Hamamatsu, Japan) in digitized form. Video signals from the small detection area ($7 \times 9\,mm^2$) of this camera were fed into an image processor (Argus-10, Hamamatsu Photonics) linked to a personal computer (PC-9821, NEC; Tokyo, Japan), which enhanced the contrast and increased the electrical magnification of the video images by ×20. The total magnification thus reached was 20000 times ($100 \times 2.5 \times 4.0 \times 20$). With this microscope, the movements of intracellular organelles/granules and gradual changes occurring within the cells could be visualized. All images obtained were recorded once on videotape for storage employing a video cassette recorder (EDV-9000B, SONY EDBeta; Sony; Tokyo, Japan), which was further connected to a High 8 video cassette recorder (EV-NS9000, SONY Hi-8), a video editor (RM-E1000, SONY) for tape editing, and a video printer (VP-9000, FUJIX; Tokyo, Japan) for immediate printing out of cut pictures if necessary. To analyze the interactions between zymosan-A and the microglia, and the process of phagocytosis at a high spatial and temporal resolution, the selected part of the video tape that was of specific interest was played back and fed into an image analyzer (Avio Excel, Nippon Avionics; Tokyo, Japan) coupled to the above computer (PC-9821, NEC). This procedure was found to be very convenient for evaluating sequential changes in nuclear size. The system was equipped with a 24-Mbyte RAM disc available for rapid storage and random access of 96 image frames. When successive frames separated by a short time delay were recorded and the resultant videotape was played at normal speed, a slow process could be speeded up by time-lapse display, and rapid processes could be contrasted.

Fifteen experiments were carried out using microglia which were obtained from the brains of newborn rats employing the technique developed at the National Institute of Neuroscience (through the courtesy of Dr. S. Kohsaka [3]). The identity of microglia thus isolated by this technique was confirmed by us on the basis of a strong phagocytic activity for Latex beads, and immunohistochemical positivity for ED1. After 7–22 days of cell culture, several round and flattened cells came to be observed on the surface of the astroglial cell layer [2]. The microglia were collected and seeded on to a coverslip for the present experiments, after confirmation of their phagocytic capability as evaluated with Latex beads. Based on distinctive morphological features, the microglia could be identified in living cultures where they were monitored using the VEC-DIC microscope. This technique revealed extremely dynamic features in the cellular plasticity and motility of the microglia. The microglia on the coverslip placed in the observation chamber were continuously superfused with DMEM, and their images were continuously stored on time-lapse videotape using the VEC-DIC microscope without a UV-cut filter. During continuous observation, zymosan-A [4] was directly, but gently, introduced from the top of the observation chamber. The magnitude of the morphological changes in the microglia before, during and after phagocytosis of zymosan-A was measured chronologically as the short axis (S), long axis (L), circumferential length (CL), and area (A) of the nucleus, or cell body.

Results

General Observations on Microglia

Cultured microglia were found to exist in four predominant forms: ramified, reactive, villous, and ameboid. The ramified or process-bearing form of microglia, which would be present in a "quiescent" state in mature brain tissue, was occasionally seen on the coverslip when microglia were seeded 2 or more days before carrying out the observations. The processes sometimes achieved a length of more than 50 μm. The ramified microglia reverted to a reative form exhibiting rapid alterations in the size and shape of their cell body with an associated retraction of processes. The reactive form of microglia was most commonly observed in the co-cultured astroglial cell layer. Such cells assumed a round flat outline with wide thin transparent lamellipodia, which were constantly being extended and retracted, thus continuously changing shape. As reported previously [2], the reactive microglia occasionally pivoted, circled, crawled and marched in one direction, utilizing thin transparent lamellipodia only as feet for slow movement, as sails to fly and glide, and as arms to swim when shear was imparted to the medium solution. However, during their movements, the microglia maintained a round cell shape with no evident agitation of the intracellular granules, and no polarization of the cell body, which formed a striking contrast to the ameboid movements of, for example, polymorphonuclear leukocytes. The diameter of the round microglia without spreading of lamellipodia was $9.03 \pm 1.06\,\mu$m (mean ± SD, $n = 33$), ranging from 7.03 to 10.36 μm. The microglia changed their shape depending on the environmental conditions to which they were exposed.

Microglia suspended in the medium after separation from astroglia appeared as an elastic spherical ball either with or without abundant filopodia (villous form). The filopodia were composed of a loose thin tube of membrane containing cytoplasm, presumably underlain and/or supported by invisible actin filaments. The filopodia moved constantly as they rapidly protruded, retracted, and fluctuated dynamically, thus growing or shrinking in length. The fine extension of the filopodia which constituted a narrow thread, occasionally attained a length of more than 10 μm. When floating villous microglia landed on the cover glass, they spread and underwent rapid remodeling into polymorphous thin flat ameboid cells surrounded by lamellipodia, with a peripheral sheet of cell membrane, increasing the cell diameter more than three-fold. The microglia appeared to become active, with increasing granules or vesicles in the cytoplasm, the majority of which were constantly gliding or traveling in various directions and over various distances. The most conspicuous vesicles were well-demarcated dark ones which were characterized by frequent alterations in their position, tending to lineup in the peripheral part of the cell body.

Interactions of Microglia with Zymosan-A and Phagocytosis

Both the villous and ameboid microglia exhibited phagocytosis. When floating villous microglia came in contact with part of the zymosan-A at the filopodia, the microglial cell began engulfment immediately without any remodeling of shape (Fig. 1). The cell utilized its filopodia skillfully to carry the zymosan-A to the cell body for phagocytosis. In some cases, the floating villous microglia, upon recognition of zymosan-A, spread rapidly and displayed a quick morphological change to an ameboid form within 10 s (Fig. 2). The ameboid microglia rapidly covered the zymosan-A with a sheet-like lamellipodia. It was barely visible how the zymosan-A was engulfed and internalized within lamellipodia. The intracellular granules appeared to be agitated and degranulated vigorously. In ameboid microglia which had already become ameboid and flattened on the growth surface, phagocytosis occurred in a slightly different manner. When aggregates of zymosan-A approached and touched the marginal portion of the microglial cell body, phagocytosis took place only at the peripheral lamellipodia. The cell body remained unchanged in shape with no granular agitation. Part of the zymosan-A was engulfed by the lamellipodia, although due to the thinness and transparency of the lamellipodia, it was quite difficult to ascertain just how much zymosan-A was actually internalized into the lamellipodial sack. A comparison of the above two types of phagocytosis suggested that it was the microglia that attacked zymosan-A in the former case, whereas in the latter, it was the zymosan-A that actively approached and became willingly engulfed by the ameboid microglia which spread out on the suface before and during phagocytosis. There must be an unknown physical force between the interaction of the cell and zymosan-A.

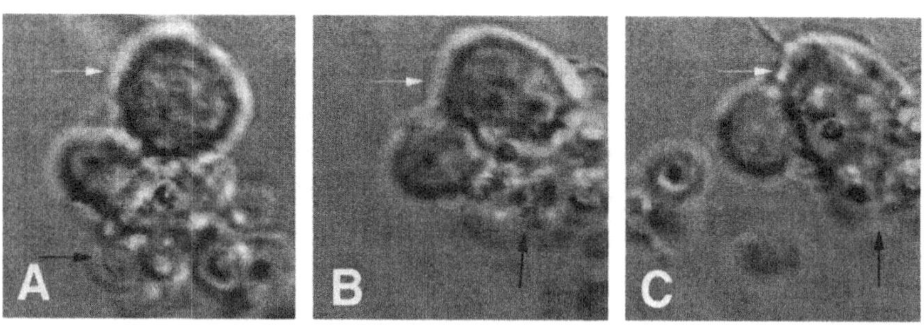

Fig. 1A–C. Phagocytosis by an villous microglial cell which engulfs zymosan-A; *white arrow* indicates microglia and *black arrow* zymosan-A

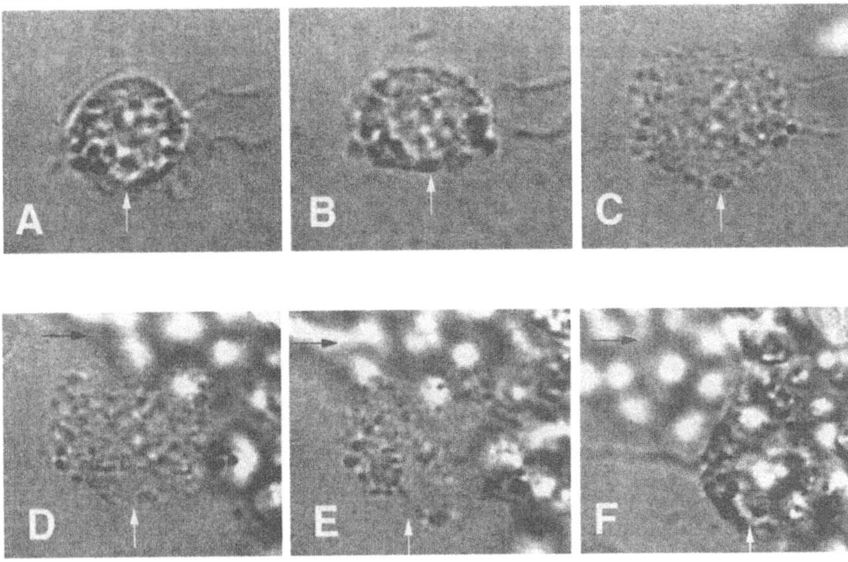

Fig. 2A–F. Swift remodeling of an villous microglial cell (time lapse of 10 s each) (*white arrow* indicates microglia and *black arrow* zymosan-A). The floating microglia upon recognition of zymosan-A spread rapidly over the zymosan-A with a sheet-like lamellipodia. Note the abundant indigestible zymosan-A within the cell body (**F**)

Cell Death and Nuclear Changes

Approximately 10–30 min after contact of zymosan-A with microglia, phagocytosis of either type induced zeiosis or blebbing of the cell membrane. The blebs joined together to form a large balloon which sooner or later burst, and the membrane became invisible. A denuded nucleus remained, which was almost unchanged in size but apparently lost its plasticity and vigor, displaying a stiff irregular appearance. The average dimensional changes in five cases where the nuclear size could be measured were 7.45, 7.32, 7.25, and 7.01 μm for the long axis, and 6.53, 6.13, 6.81, and 6.55 μm for the short axis in the control, at the time before bursting, after bursting, and at 10 min after bursting, respectively (Table 1). The cell body changes in terms of the circumferential length were: 84.38 μm in the control and 141.71 μm at the time of bursting ($p < 0.05$), and those in the area of the cell body were 36.43 μm^2 in the control and 47.92 μm^2 at the time of bursting ($p < 0.05$). Figure 3 shows several dead nuclei after zymosan-A phagocytosis. Although no fragmentation of the nucleus was detected, chromatin condensation was evident in Fig. 3C–E, and H. Fig. 3K depicts a polymorphonuclear leukocyte in a state of coagulation necrosis which had adhered to a pig brain endothelial cell for 3 h, and Fig. 3L shows a necrotic neuron after 24-h asphyxiation; these are presented for a comparison of the nuclear changes.

Table 1. Microglial cell dimensional changes associated with cell death after phagocytosis

	Control	Before bursting	After bursting	10 min after bursting
LAN	7.45 ± 1.85	7.32 ± 1.21	7.25 ± 2.39	7.01 ± 1.88
SAN	6.53 ± 2.18	6.13 ± 1.67	6.81 ± 1.44	6.55 ± 1.16
CLCB	84.38 ± 37.88	141.71 ± 37.81*	–	–
ACB	36.43 ± 6.73	47.92 ± 7.04*	–	–

LAN, Long axis of the nucleus; SAN, short axis of the nucleus; CLCB, circumferential length of the cell body; ACB, area of the cell body; –: cell had burst.
$^*p < 0.05$.

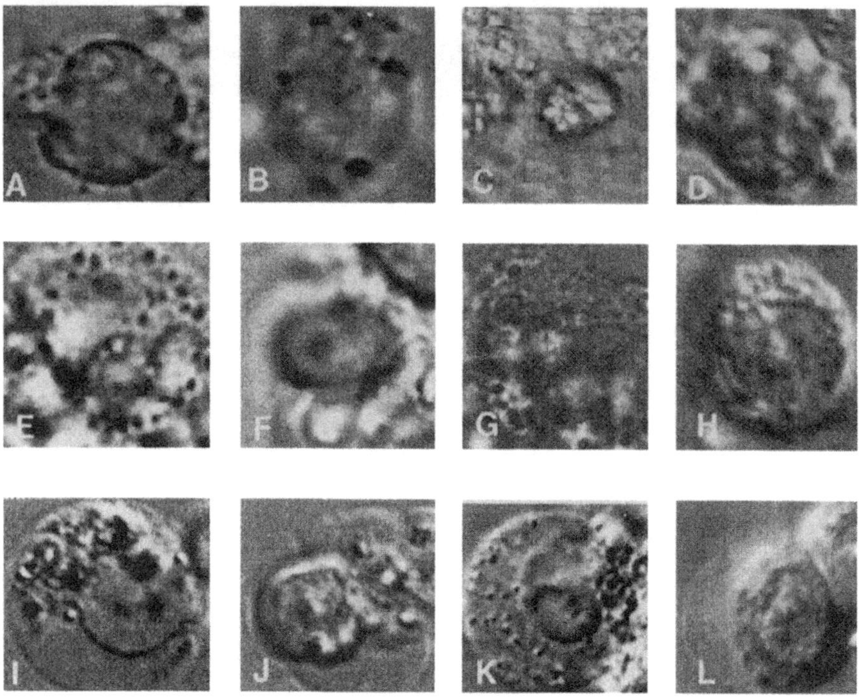

Fig. 3A–L. Dead nuclei after zymosan-A phagocytosis (see text for further explanation). **A–J**, microglia; **K**, polymorphonuclear leukocyte; and **L**, neuron

Discussion

It is conceivable that microglia may be fated to die when they have completed their work of phagocytosis. Such cell death could be apoptotic. The fate of the apoptotic cell is to be phagocytosed by one of its nearby siblings or by a professional phagocyte. In general, apoptosis has been depicted as beginning with cell shrinkage; the cytoplasm becomes crowded, but with normal intracellular organelles, and the nucleus soon develops extremely condensed chromatin that appears to associate with the nuclear envelope. The plasma membrane becomes active and ruffled in the so-called process of zeiosis [4]. The microglial cell death described above, with loss of membrane integrity resulting in cell swelling and lysis and lysosomal leakage after phagocytosis of zymosan-A, favors necrosis. In our study, zymosan-A apparently triggered self-inflicted membrane disruption of the microglia, possibly by disrupting the ionic balance, although the changes were perhaps enhanced by the noxious stimulus of UV light. Due to a cascade of thermodynamic potential, the damage to the membrane integrity may result in a rush of sodium ions into the cell interior accompanied by water. Thus, under these experimental conditions, activated microglia ultimately succumb to cell death as a result of cell swelling and, therefore, presumably necrosis. However, the nuclear change of chromatin condensation with cell shrinkage and swelling of the nucleus observed in some cases, suggests that the death could involve apoptosis. Indeed, the cell death described above appeared to have components of both necrotic and apoptotic cell death. Terakawa (personal communication: Prof. Susumu Terakawa, Photon Medical Research Center, Hamamatsu University School of Medicine, Hamamatsu 431-31, Japan) has coined such cell death "apo-necrosis".

Acknowledgments. We wish to thank Dr. Shinichi Kohsaka, Department of Neurochemistry, National Institute of Neuroscience, Tokyo 187, Japan, for providing us with microglia and for instructing us in his group's technique for collecting microglia. The research reported here was supported by the Japan Allergy Foundation and Hoechst Japan.

References

1. Giulian D (1987) Ameboid microglia as effectors of inflammation in the central nervous system. J Neurosci Res 18: 155–171
2. Haapaniemi H, Tomita M, Tanahashi N, Takeda H, Yokoyama M, Fukuuchi Y (1995) Non-amoeboid locomotion of cultured microglia obtained from newborn rat brain. Neurosci Lett 193: 121–124
3. Nakajima K, Hamanoue M, Shimojo M, Takei N, Kohsaka S (1989) Characterization of microglia isolated from a primary culture of embryonic rat brain by a simplified method. Biomed Res 10 [Suppl 3]: 411–423
4. Squier MKT, Sehnert AJ, Cohen J (1995) Apoptosis in leukocytes. J Leukoc Biol 57: 2–10
5. Terakawa S, Fan IH, Kumakura K, Ohara-Imaizumi M (1991) Quantitative analysis of exocytosis directly visualized in bovine chromaffin cells by videoenhanced light microscopy. Neurosci Lett 123: 82–86
6. Tomita M, Fukuuchi Y, Tanahashi N, Kobari M, Takeda H, Yokoyama M, Ito D, Terakawa S (1996) Swift transformation and locomotion of PMNL and microglia as observed by VEC-DIC microscopy (video microscopy). Keio J Med 45(3): 213–224

Mitochondrial Respiratory Enzyme Activity and the Maturation Phenomenon in Cerebral Ischemia

T. Kuroiwa[1], U. Ito[2], and R. Okeda[1]

Summary

We examined regional differences in the activity of a mitochondrial respiratory enzyme, succinic dehydrogenase (SDH), using a quantitative imaging method in normal brains and in brains subjected to transient global or permanent focal ischemia. SDH activity in the pyramidal cell layer of the CA 1 sector (Sommer's sector) in normal gerbils was lower than in any other subfield in the hippocampus. After transient global ischemia induced by bilateral common carotid artery occlusion for 5 min, SDH activity in the CA 1 sector remained unchanged. Histologically, selective neuronal necrosis was observed in the pyramidal cell layer of the CA 1 sector 2 days after ischemia. In focal ischemia induced by permanent occlusion of the left common carotid artery in gerbils, a patchy area of reduced SDH activity appeared at the ischemic center 3 h after occlusion. This area occurred irrespective of the pre-ischemic SDH activity level and enlarged to include the ischemic periphery (penumbra) over the following 3 h. Both delayed neuronal death of CA 1 neurons and enlargement of the infarction to include the penumbra are slowly developing ischemic injuries. However, this study indicates that the effect of energy metabolism and its impairment on these phenomena is different.

Introduction

Energy metabolism and its impairment are closely associated with ischemic injury, both in terms of selective vulnerability after global ischemia and in the enlargement of cerebral infarction to include the penumbra. After transient global ischemia, programmed cell death – an energy-requiring process – takes place in the postischemic CA 1 sector [6, 7]. In the normal brain, local cerebral glucose utilization in the pyramidal cell layer of the CA 1 sector is lower than in the other sectors and layers of the hippocampus [21]. Several mitochondrial enzymes are sporadically distributed in the hippocampus [4, 5, 16], for example, the level of mRNA for cytochrome C oxidase

[1] Department of Neuropathology, Medical Research Institute, Tokyo Medical and Dental University, Yushima 1-5-45, Bunkyo-ku, Tokyo 113, Japan
[2] Department of Neurosurgery, Musashino Red Cross Hospital, 1-26-1 Kyonan-cho, Musashino-shi, Tokyo 180, Japan

Maturation Phenomenon in Cerebral Ischemia II
U. Ito et al. (Eds.)

decreases during early recirculation in the CA 1 sector [1]. Thus, levels of indices for tissue energy metabolism are not consistent and change frequently in both the normal and postischemic hippocampus.

In focal cerebral ischemia, the gradual encroachment of ischemic infarction into the surrounding penumbra is closely associated with energy failure in the ischemic tissue [10, 18]. In this study, we investigated the difference between delayed neuronal death and enlargement of the infarction to include the penumbra, processes which both involve delay in the development of ischemic change, using quantitative imaging of the activity of a mitochondrial respiratory enzyme, succinic dehydrogenase (SDH).

Materials and Methods

Quantitative SDH Activity Imaging

For quantitative imaging of SDH, coronal sections of gerbil brain were incubated in 1% 2,3,5 triphenyl tetrazolium chloride (TTC) in Dulbecco's tissue culture medium at 35°C for 6 min. Regional changes in the optical density (OD) of the coronal surface due to the accumulation of formazan produced from TTC by SDH were recorded, employing color standards at 1, 3 and 6 min of immersion, and using a charge-coupled device (CCD) camera. The formazan content was then calculated according to the correlation between the regional tissue formazan content and the corresponding OD. Regional SDH activity was calculated from the initial rate of formazan production and displayed using semi-color coding. The correlation between SDH activities measured by this method [10, 12] and by a standard biochemical method [13] was examined as follows. Coronal sections of control brains were prepared for quantitative imaging as described above, and adjacent sections were frozen and sampled from the mirror surfaces at various regions of interest. The correlation between the data obtained by these two independent methods was evaluated statistically.

Laboratory Preparation

To obtain control values for SDH activity, brains of untreated gerbils were removed under ketamine anesthesia (intramuscular injection, 50 mg/kg) and placed in a moist chamber maintained at 0°C. Coronal sections from several levels were prepared for quantitative imaging [14]. After this procedure, the coronal sections were immersion-fixed with 4% buffered formalin, embedded in paraffin and stained with hematoxylin & eosin and cresyl violet–luxol fast blue for histological evaluation.

Transient global ischemia was induced in gerbils by bilateral common carotid artery occlusion for 5 min while maintaining the body temperature at 37°C under ketamine anesthesia. Animals were sacrificed 12 h or 2 days later. Focal ischemia was induced in gerbils by left common carotid artery occlusion with a microvascular clip through a ventral midline incision in the neck under ketamine anesthesia [8]. Only typically ischemia-sensitive animals were selected for subsequent study. Animals in the ischemic groups were sacrificed 1 h, 3 h or 6 h after the start of occlusion. The brains of both groups were removed immediately and coronal sections of forebrain were prepared for the imaging procedure described above.

Data were analyzed using one-way analysis of variance and Fisher's protected least significant difference procedure for multiple comparisons.

Results and Conclusion

Quantitative SDH Activity Imaging

Regional SDH activity was quantified by densitometric measurement of the rate of formazan production [10, 12]. SDH activity measured by this method showed significant linear correlation with results obtained using the standard biochemical method ($R^2 = 0.956$) [13]. The distribution of SDH in the normal brain has been studied in different animal species, but regional differences in the hippocampus have not been reported previously. In previous studies, the distribution of SDH was evaluated after the enzymatic staining reaction had reached a plateau. By contrast, we measured the initial rate of formazan production, and detected a small difference in SDH activity which was not found after coloration reached plateau levels. Using this method in experimental focal cerebral ischemia in the cat, we previously detected a decrease in SDH activity to less than 15% of the control value. This was detectable in the center of the ischemic area as early as 3h after the onset of ischemia [11].

Selective Vulnerability of CA 1 Neurons

In control gerbils, SDH activity in the pyramidal cell layer of the CA 1 sector was lower than in any other subfield of the hippocampus. SDH activity in the oriens layer, stratum radiatum and lacunosum molecular layer of the CA 1 sector was lower than in the corresponding layers of the CA 2 and CA 3 sectors. Twelve hours after transient global ischemia induced by bilateral common carotid artery occlusion over 5min, SDH activity showed a very mild decrease in the CA 1 sector as shown in Fig. 1. SDH activity in all layers of the CA 1 sector was not significantly lower than control values. SDH activity remained unchanged in the CA 1 sector 48h after transient ischemia. Histologically, selective neuronal necrosis was observed in the pyramidal cell layer of the CA 1 sector 48h after ischemia.

Ischemic Penumbra in Focal Ischemia

SDH activity was significantly reduced in the dorsolateral thalamus 3h after the onset of ischemia, and this region was surrounded by an area of mild SDH activity reduction (Fig. 2). The area of reduced activity was independent of the pre-ischemic SDH activity level and enlarged to include the ischemic periphery (penumbra) over the following 3h. The area showing reduced SDH activity at 3h became sharply demarcated at 6h. In the cortex, small foci of reduced SDH activity appeared at 3h, and these enlarged to involve the surrounding cortex by 6h. In the hippocampus, reduction of

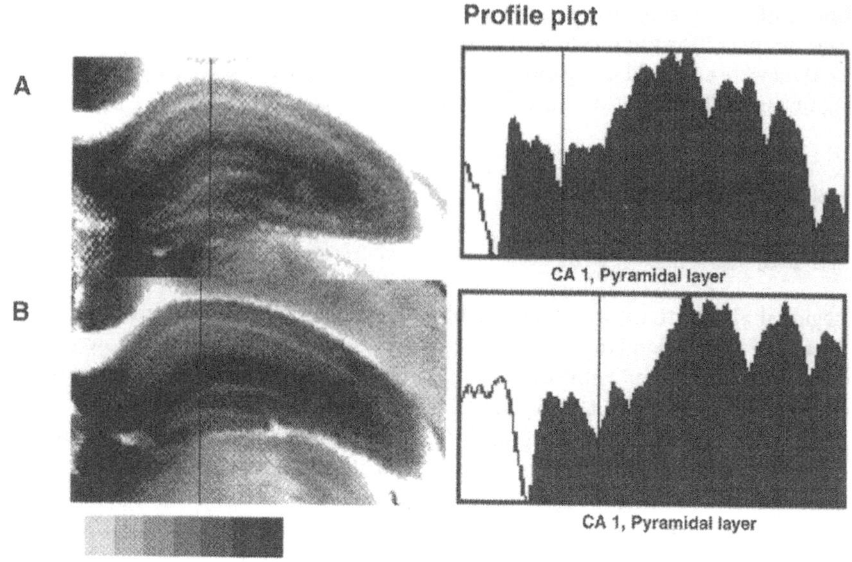

Fig. 1A,B. Quantitative succinic dehydrogenase (SDH) activity imaging (*left*) and profile plot of the CA 1 sector (*right*) of gerbil hippocampus in control animals (**A**) and 12 h after occlusion of both common carotid arteries for 5 min (**B**). Note that SDH activity was lowest in the pyramidal cell layer of the CA 1 sector in both control and post-ischemic brains

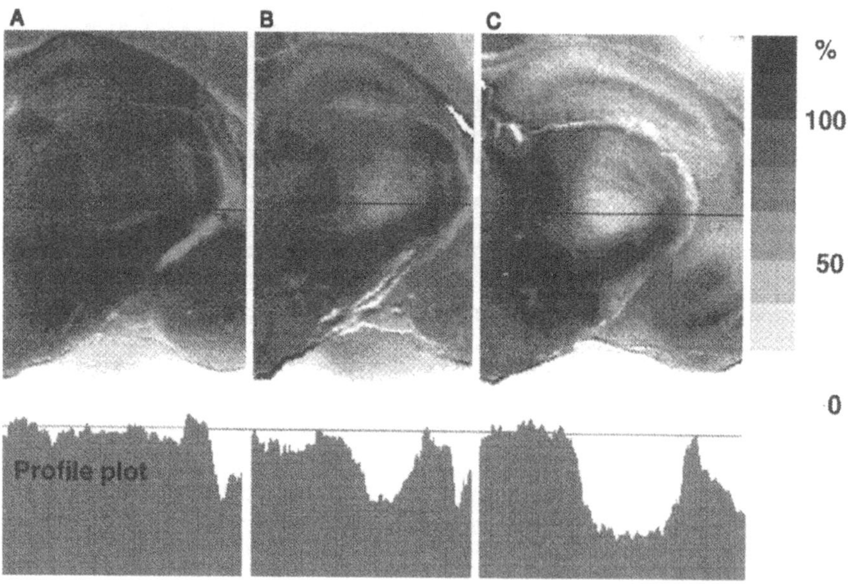

Fig. 2A–C. Quantitative succinic dehydrogenase (SDH) activity imaging (**A–C**, upper) and profile plot (**A–C**, lower) of the dorsolateral thalamus of gerbil brain in control animals (**A**) and 3 h (**B**) and 6 h (**C**) after unilateral permanent common carotid artery occlusion. Note that SDH activity was decreased in the dorsolateral thalamus. The area of marked activity reduction enlarged and the border of the surrounding area became sharp 6 h after the onset of ischemia. Infarction of the entire hippocampus became evident 6 h after the onset of permanent ischemia

SDH activity was evident by 6h. SDH activity was reduced in the entire hippocampus and was not just the CA 1 sector, as shown in Fig. 2. Mild reduction in SDH activity 3h after the onset of ischemia preceded light-microscopical changes in the ischemic tissue, whereas the severe reduction at 6h corresponded to areas of infarction.

SDH Activity and Maturation Phenomenon

This study has demonstrated that SDH activity in the pyramidal cell layer of the CA 1 sector is lower than in other layers of the CA 1 sector and is also lower than in the pyramidal cell layer of the CA 2 and CA 3 sectors. A similar pattern has been observed for local cerebral glucose utilization and intensity of immunoreactivity for the pyruvate dehydrogenase complex, a mitochondrial enzyme of the tricarboxylic acid (TCA) cycle [4, 16] in the hippocampus. Thus, low values for several indices of cellular energy metabolism are observed in the pyramidal cell layer of the CA 1 sector. Glucose utilization is not reduced in the CA 1 sector during early reperfusion after transient forebrain ischemia, which is not the case in regions resistant to ischemic damage [17]. Inhibition of postischemic excitatory input to the CA 1 sector reduces glucose utilization [9] and prevents postischemic neuronal death in the CA 1 sector [20]. These observations indicate that a low capacity for mitochondrial respiration may become a critical factor triggering delayed neuronal death in the CA 1 sector during early reperfusion. The low level of SDH activity seen in the normal pyramidal cell layer of the CA 1 sector remained unchanged during the onset of delayed neuronal death. This finding is quite different from the situation in acute cerebral infarction, in which SDH activity is rapidly lost from the perfusion area of the occluded artery, irrespective of the normal distribution of SDH in controls. Levels of ATP, ADP and phosphocreatinine in the CA 1 sector recover rapidly after transient global ischemia and are equivalent to preischemic levels during early reperfusion [2]. Thus, the process leading to delayed neuronal death is not a passive consequence of cellular energy failure, but is probably an energy-requiring process. One of the possible mechanisms is apoptosis [6, 7]. The present study indicates that the process leading to delayed neuronal death is triggered in an area with relatively poor cellular energy metabolism reserves, and proceeds without energy failure.

The penumbra [3] has been defined as an area with blood flow ranging between the thresholds for transmitter and membrane failure, in which neuronal activity is suppressed but the membrane function essential for cell survival is maintained. However, recent studies have indicated that the ischemic penumbra is not necessarily electrically silent, nor is the ion homeostasis and energy metabolism necessarily preserved [15, 18, 19].

The present study has also demonstrated a gradual reduction of SDH activity in the zone surrounding an area of cerebral ischemia over the 6h after the onset of ischemia. Collectively, these findings indicate that impairment of energy metabolism during early ischemia plays a central role in the recruitment of the peripheral zone into the infarction process.

Both delayed neuronal death and recruitment of the penumbra into the infarction process are phenomena in which slower development of ischemic tissue injury occurs, as compared to the acute process of infarction in the ischemic center. However, the results presented here indicate that the role of energy metabolism in these two events is different.

Acknowledgments. We thank Drs. M Ueki (Department of Anesthesiology, Tokyo Medical and Dental University), T Yamaguchi (Department of Biochemical Genetics, Tokyo Medical and Dental University) and Z Czernicki (Department of Neurosurgery, Tokyo Medical and Dental University) for their participation in parts of this project.

References

1. Abe K, Kawagoe J, Aoki M, Kogure K (1993) Changes of mitochondrial DNA and heat shock protein gene expression in gerbil hippocampus after transient forebrain ischemia. J Cereb Blood Flow Metabol 13: 773–780
2. Arai H, Passoneau JV, Lust D (1986) Energy metabolism in delayed neuronal death of CA 1 neurons of the hippocampus following transient ischemia in the gerbil. Metab Brain Dis 1: 263–278
3. Astrup J, Siesjö BK, Symon L (1981) Threshold in cerebral ischemia. The ischemic penumbra. Stroke 12: 726–730
4. Bagley PR, Tucker SP, Nolan C, Lindsay JG, Davies A, Baldwin SA, Cremer JE, Cunningham VJ (1989) Anatomical mapping of glucose transporter protein and pyruvate dehydrogenase in rat brain: an immunogold study. Brain Res 499: 214–224
5. Calingansan NY, Baker H, Sheu KR, Gibson GE (1994) Distribution of the alpha-ketoglutarate dehydrogenase complex in rat brain. J Comp Neurol 346: 461–479
6. Goto K, Ishige A, Seiguchi K, Lizuka S, Sugimoto A, Yuzurihara M, Aburada M, Hosoya E, Kogure K (1990) Effects of cycloheximide on delayed neuronal death in rat hippocampus. Brain Res 534: 299–302
7. Heron A, Pollard H, Dessi F, Moreau J, Lasbennes F, Ben-Ari Y, Charriaut-Marlangue C (1993) Regional variability in DNA fragmentation after global ischemia evidenced by combined histological and gel electrophoresis observations in the rat brain. J Neurochem 61: 1973–1976
8. Ito U, Spatz M, Walker JT, Klatzo I (1975) Experimental cerebral ischemia in Mongolian gerbils I Light microscopical observations. Acta Neuropathol (Berl) 32: 209–223
9. Jorgensen MB, Wright DC, Diemer NH (1990) Postischemic glucose metabolism is modified in the hippocampal CA 1 region depleted of excitatory input or pyramidal cells. J Cereb Blood Flow Metab 10: 243–251
10. Kuroiwa T, Czernicki Z, Ueki M, Kaneko H, Okeda R (1993) Quantitative imaging of succinic dehydrogenase activity in gerbil brain after unilateral common carotid artery occlusion. J Cereb Blood Flow Metab [Suppl] 13: 738
11. Kuroiwa T, Ueki M, Chen Q, Suemasu H, Taniguchi I, Okeda R (1994) Biomechanical characteristics of brain edema. Acta Neurochir (Wien) [Suppl] 60: 158–161
12. Kuroiwa T, Terakado M, Yamaguchi T, Endo S, Ueki M, Okeda R (1996) The pyramidal cell layer of sector CA 1 shows the lowest hippocampal succinic dehydrogenase activity in normal and postischemic gerbils. Neurosci Lett 206: 1–4
13. Lippold HJ (1982) Quantitative SD histochemistry. Histochemistry 76: 381–405
14. Loskota WJ, Lomax P, Verity MA (1974) A stereotaxic atlas of the mongolian gerbil brain. Ann Arbor Science, Ann Arbor, Michigan, pp 1–567
15. Mies G, Ishimaru S, Xie Y, Seo K, Hossmann KA (1991) Ischemic thresholds of cerebral protein synthesis and energy state following middle cerebral artery occlusion in rat. J Cereb Blood Flow Metab 11: 753–761
16. Milner TA, Aoki C, Rex Sheu K-F, Blass JP, Pickel VM (1987) Light microscopic immunocytochemical localization of pyruvate dehydrogenase complex in rat brain. J Neurosci 7: 3171–3190
17. Pulsinelli WA, Lewy DE, Duffy TE (1982) Regional cerebral blood flow and glucose metabolism following transient forebrain ischemia Ann Neurol 11: 499–509
18. Selman WR, Lust D (1991) Pathophysiology and treatment of focal cerebral ischemia. Neurosurgeons 10: 18–27

19. Siesjö BK (1992) Pathophysiology and treatment of focal cerebral ischemia Part I: Pathophysiology. J Neurosurg 77: 169–184
20. Wieloch T, Lindvall O, Bromqvist P, Gage FH (1985) Evidence for amelioration of ischemic neuronal damage in the hippocampal formation by lesions of the perforant path. Neurol Res 7: 24–26
21. Wree A, Schleicher A, Zilles K, Beck TV (1988) Local cerebral glucose utilization in the Ammon's horn and dentate gyrus of the rat brain. Histochemistry 88: 415–426

Maturation of Cell Damage After Repetitive Hypoxia–Ischemia in the Rat

H.C. Klein[1], K.G. Go[2], and J. Korf[1]

Summary

Rats with unilateral carotid ligation were subjected to 10-min periods of hypoxia (10% O_2) with 20-min intervals of normoxia (30% O_2). Development of cytotoxic edema during hypoxia was monitored by the elevation of striatal impedance. After one hypoxic episode the impedance change was always reversible, but after subsequent hypoxic episodes the impedance change progressively became irreversible. Compared to the animals with reversible change of impedance, those with irreversible change had a more elevated striatal Na^+/K^+-ratio, and more extensive histological damage as assessed by cellular silver uptake. But even in the animals with reversible impedance change histological damage matured, becoming more extensive from 4h to 24h of survival. Hypothermia (striatal temperature of 28°C) and 48h of food deprivation proved to have a protective effect in delaying the onset of irreversible striatal impedance change.

Introduction

There are clinical conditions in which the occurrence of repetitive hypoxia–ischemia is not infrequent. These are situations in which the maintenance of cerebral perfusion pressure is critical or in which the cerebrovascular responses are impaired. Cerebral perfusion is endangered in conditions of unstable arterial pressure, vascular stenosis or spasm, and in the presence of mass lesions causing elevated intracranial pressure. Impaired cerebrovascular reactivity to arterial hypotension or to hypoxia has been demonstrated after head injury. Furthermore, during vascular surgery repetitive ischemia may occur during repetitive occlusion of feeding vessels. Also, the risk of repetitive ischemia is not inconceivable in patients with heart failure or serious cardiac arhythmia.

As a simple model that has a great resemblance to the clinical situation we have chosen the Levine model of unilateral carotid ligation in the rat with repetitive hypoxia. In previous studies we have demonstrated that beside hypoxia, ischemia is

[1] Department of Biological Psychiatry, University Hospital Groningen, Hanseplein 1, P.O.B. 30.001, Groningen 9700 RB, The Netherlands

[2] Department of Neurosurgery, University Hospital Groningen, Hanseplein 1, P.O.B. 30.001, Groningen 9700 RB, The Netherlands

Maturation Phenomenon in Cerebral Ischemia II
U. Ito et al. (Eds.)

greatly involved in the Levine model, since during the hypoxic episode there is a considerable fall of arterial blood pressure, which is presumably of cardiac origin, and which is accompanied by a consequent decrease of cerebral blood flow [3]. We have chosen 20-min intervals between the hypoxic–ischemic episodes, while Hossmann et al. [4] have investigated ischemia at hourly intervals in cats. In addition, the model was employed to assess the effects of fasting and hypothermia.

Materials and Methods

Male Wistar rats (240 g) were anesthetized with pentobarbital (40 mg/kg i.p.). Two tungsten impedance electrodes 0.2 mm in diameter (insulated with epoxy resin, except 0.5 mm at the tip) were stereotactically implanted, 1 mm apart bilaterally into the striatum, and fixed to the skull. Impedance was measured with an alternating (10 kH) current of 20 μA. The left common carotid artery was distally ligated and proximally cannulated for measurement of arterial blood pressure and monitoring of blood gases and glucose. To prevent collateral circulation the left internal carotid artery was also ligated. The animals were ventilated through a tracheostomy with 70% N_2O and 30% O_2. Rectal temperature was continuously monitored and kept at 37°C by a heating pad. A number of hypoxic episodes were induced by ventilation with 10% O_2 for 10 min, and repeated after intervals of 20 min. The animals were studied after survival periods of 4 h and 24 h. Some animals were used for the assay of whole tissue Na^+ and K^+ contents by flame emission spectroscopy. In other animals, histology was performed after perfusion fixation with 4% paraformaldehyde solution. To monitor cellular damage, brain sections were stained with a modified silver impregnation technique [1] and photographed in dark field illumination. Blood glucose was measured by spectrophotometry.

In some animals microdialysis was performed by means of a dialysis tube (100 μm diameter; permeability to a molecular mass up to 10 kDa), which had been inserted transversely through the head at a level of 0.5 mm below the impedance electrodes on the previous day. Perfusion was performed with a fluid containing 120 m*M* NaCl, 15 m*M* $NaHCO_3$, 5 m*M* KCl, 1.5 m*M* $CaCl_2$, 1 m*M* $MgSO_4$, pH 7.4 at a rate of 7 μl/min. Glucose was assayed on-line by addition of 1.6 U of hexokinase, 0.8 U glucose-6-P dehydrogenase, 0.4 m*M* ATP and 3 MM nicotinamide adenine dinucleotide phosphate (NADP) in a Tris-buffer, pH 8; the endproduct nicotinamide adenine dinucleotide phosphate, reduced (NADPH) was measured by fluorescence spectrometry.

To assess the influence of fasting, a group of animals received no food for 48 h prior to the experiment. The effect of hypothermia was evaluated in another group of animals, which had a thermocouple of 0.25 mm diameter (Ellab, Copenhagen, Denmark) implanted into the left striatum beside the impedance electrode, and then subjected to local cooling of the head by application of ice cubes (to a striatal temperature of 28°C), while the body's temperature was maintained at 37°C by a heating pad.

Results and Comments

During normoxia the mean arterial blood pressure was 128.1 ± 2.3 mm Hg, versus 50.9 ± 2.8 mm Hg during hypoxia. The arterial oxygen pressure amounted to 21.36 kPa during normoxia, versus 4.56 ± 0.17 kPa during hypoxia, while the arterial CO_2 pressure was 5.96 ± 0.58 kPa during normoxia and 5.48 ± 0.39 kPa during hypoxia, with a pH of 7.22 ± 0.03 during normoxia and 7.13 ± 0.01 during hypoxia, demonstrating hypoxemia as well as arterial hypotension during the hypoxic episodes.

Since low-frequency alternating current is conducted through the extracellular space, increase of brain tissue impedance in cerebral hypoxia–ischemia indicates shrinking of cerebral extracellular space due to cell swelling of cytotoxic edema [2, 5]. Following one hypoxic period the change of striatal impedance was reversible in all animals; after two hypoxic periods the impedance elevation was irreversible in 10%, after three to five hypoxic periods in 25%, and after five periods in 76% of the animals (Fig. 1), illustrating the cumulative effect of repetitive insults.

In animals with irreversible change of striatal impedance following five hypoxic episodes and 3 h of survival, the striatal Na^+/K^+-ratio amounted to 1.14 ± 0.18 ($p < 0.001$) in the ipsilateral striatum and 0.44 ± 0.06 in the contralateral striatum, whereas in animals with reversible impedance changes it was 0.43 ± 0.03 ipsilaterally, and 0.36 ± 0.01 contralaterally.

In the animals with reversible impedance elevation, cellular damage after 4 h of survival was only present in certain sensitive areas, such as the substantia nigra, but not the hippocampus, whereas in the animals exhibiting irreversible impedance elevation, cellular damage was more extensive and more severe, including the ipsilateral striatum, cerebral cortex and thalamus (Fig. 2). But even in the animals with reversible impedance elevation, after 24 h of survival, the silver uptake had become significantly more extensive than after 4 h, including the hippocampus and demonstrating the maturation of cellular damage.

In the microdialysate, glucose concentration decreased during the hypoxic episodes and recovered during the normoxic intervals, illustrating the decrease of glu-

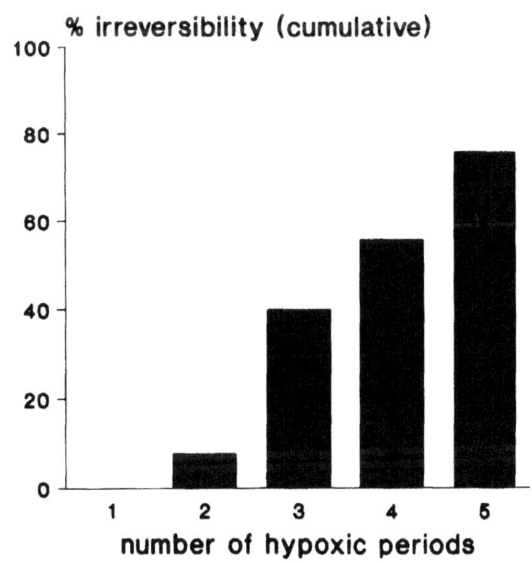

Fig. 1. Cumulative percentage of irreversibility of striatal impedance change versus the number of hypoxic episodes

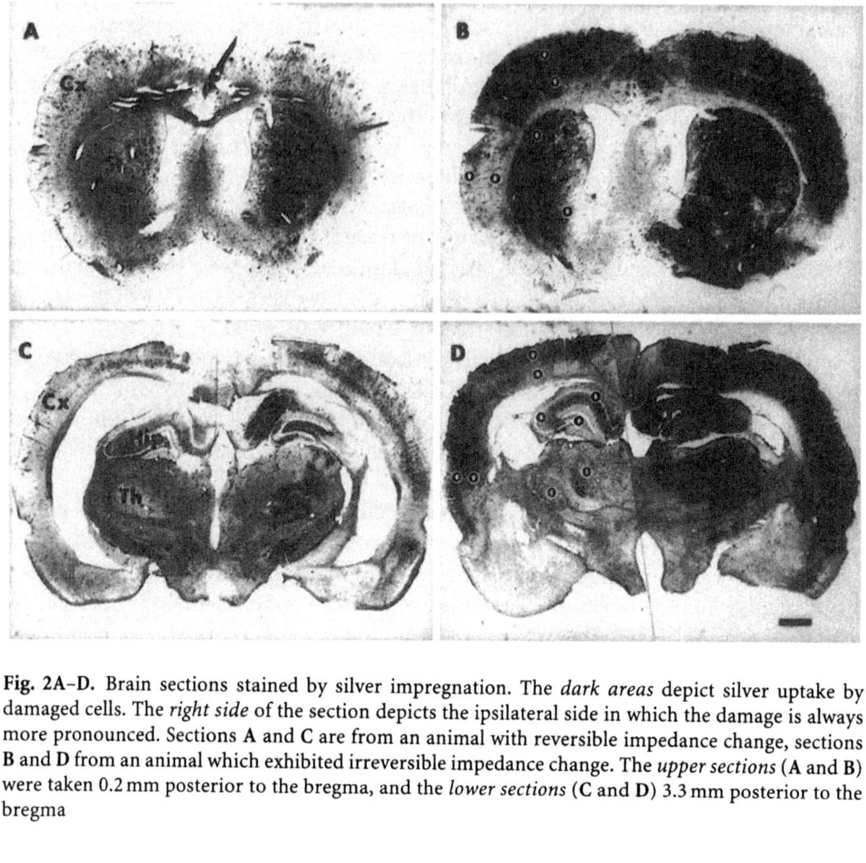

Fig. 2A–D. Brain sections stained by silver impregnation. The *dark areas* depict silver uptake by damaged cells. The *right side* of the section depicts the ipsilateral side in which the damage is always more pronounced. Sections **A** and **C** are from an animal with reversible impedance change, sections **B** and **D** from an animal which exhibited irreversible impedance change. The *upper sections* (**A** and **B**) were taken 0.2 mm posterior to the bregma, and the *lower sections* (**C** and **D**) 3.3 mm posterior to the bregma

cose supply as a result of the hypoxia-induced hypotension and reduction of blood flow, in spite of elevation of blood glucose level from 5.3 ± 0.2 m*M* before to 8.4 ± 0.7 m*M* during hypoxia. In the fasted rats blood glucose was 4.3 ± 0.1 m*M* before hypoxia, the elevation during hypoxia was attenuated (4.7 ± 0.4 m*M*); and glucose concentration was lower in the microdialysate, while recovery of glucose concentration during the normoxic intervals progressively decreased after a number of hypoxic episodes.

Whereas in normally fed and in normothermic animals reversible impedance change only occurred in the first 2.8 ± 0.1 hypoxic periods, the food-deprived rats exhibited reversibility of striatal impedance change even in 6.8 ± 0.2 hypoxic episodes, and the rats with striatal hypothermia of 28°C still had reversible impedance change in 6.4 ± 0.4 hypoxic episodes. This demonstrated the protective effects of fasting and hypothermia in delaying the onset of irreversible change of striatal impedance. Histologically, the food-deprived rats showed no cellular silver uptake in cerebral cortex, striatum, substantia nigra, and thalamus compared to the normally fed animals.

It may be concluded that seemingly "light" episodes of hypoxia-ischemia which alone do not induce irreversible damage by repetition may result in a cumulative detrimental effect, as demonstrated by the impedance changes, the increase of Na^+/K^+-ratio, and the amount of cellular damage as shown by silver uptake, while food deprivation and hypothermia were protective.

References

1. Gallyas F, Wolff J, Böttcher H, Zaborsky L (1980) A reliable and sensitive method to localize terminal degeneration and lysosomes in the central nervous system. Stain Technol 55: 299–306
2. Gazendam J, Go KG, Van der Meer J, Zuiderveen F (1979) Changes of electrical impedance in edematous cat brain during hypoxia and after intracerebral ouabain injections. Exp Neurol 55: 78–87
3. Go KG, Prenen GHM, Korf J (1988) Protective effect of fasting upon cerebral hypoxic-ischemic damage. Metab Brain Dis 3: 257–263
4. Hossmann KA, Nagashima G, Klatzo I (1990) Repetitive ischaemia of cat brain: pathophysiological observations. Neurol Res 12: 158–164
5. Van Harreveld A (1966) Brain tissue electrolytes. Butterworths, London

Synopsis and Round Table Discussion

The present Second Maturation Phenomenon Workshop highlights the distance covered since the First Workshop in the understanding and elucidation of various mechanisms and features of the phenomenon. While in the First Workshop the interpretation of various findings was frequently tied to hypothetical assumptions, in the present workshop these assumptions have substantially acquired a solid basis as the result particularly of recent advances in molecular biology. The pathological changes in cerebral ischemia are becoming increasingly associated with the interplay between various genetic expressions. Thus, at this Workshop neuronal death and recovery have repeatedly been associated with either apoptosis-related genetic expressions and with the induction of various neurotrophic factors (NtFs) shown to ameliorate ischemic injury. In this respect, the first most relevant presentation at the meeting was that of Dr. F.R. Sharp on the *evolution of genetic expressions* in hippocampus following focal and global ischemia.

His observations revealed that neurons destined to survive in the area of the penumbra are associated with upregulated HSP70 immunoreactivity, whereas neurons destined to die, such as the majority of CA1 pyramidal cells, reveal positive TUNEL staining, denoting the induction of apoptosis-related mechanisms. He also demonstrated that, in agreement with the principle of maturation phenomenon, the induction of apoptosis-related genes follows a delay that is shorter in animals exposed to more intense insults. In the following presentation, Dr. J. Ikeda's exquisite observations in vitro suggested that *involvement of apoptosis* can be assumed in the delayed type of neuronal death, whereas an acute excitotoxic neuronal death seems to fail in fulfilling the criteria for apoptosis.

For practical reasons, with regard to a potential clinical application, a large portion of the presentations and discussions were related to *protective mechanisms* which may be induced and manipulated following an ischemic insult. It is now recognized that major events following an ischemic insult consist in the release of glutamate, associated with the upregulated expression of immediate-early genes and the induction of various NtFs.

Although protective and regenerative effects of NtFs have been shown in many experimental studies in recent years, a major obstacle for their effective application in vivo has been the *impermeability of the blood-brain barrier* for these compounds. The presentation of Dr. J. Krieglstein demonstrated that NtFs such as nerve growth factor can be induced in the brain by the intraperitoneal administration of lipophilic drugs which easily cross the blood-brain barrier; this can be considered as one of the highlights of this Workshop. In this elegant study using both elenbuterol, a β_2-adrenergic receptor agonist, and selegiline, an irreversible monoamine oxidase B

Maturation Phenomenon in Cerebral Ischemia II
U. Ito et al. (Eds.)

inhibitor. Dr. Krieglstein showed that both compounds significantly increase nerve growth factor in vitro, and that both have a markedly increased protective effect against excitotoxic injury both in vitro and in vivo when rats subjected to occlusion of the middle cerebral artery are pretreated with one of these drugs. These observations may open new horizons for studies on the therapeutic administration of NtFs in various clinical conditions associated with brain pathology.

The other new approach for the prevention or amelioration of ischemic damage is related to the transneuronal activation of protective and regenerative responses in regions affected by ischemia. Although *transneuronal induction of tolerance* refers only to areas previously unaffected by ischemia, as demonstrated at this Workshop, it is conceivable that in the ischemic territory an amelioration of penumbral injury may also be achieved by transneuronal stimulation from distant but easily accessible sites which themselves remain free of any appreciable damage to brain tissue. Otherwise an exploration of transneuronally induced tolerance in remote areas, which are "pure" for being unaffected directly by an ischemic insult, may be particularly promising for the analysis of events and factors specifically related to tolerance and for the eventual identification of "magic bullets" directly responsible for the induction of tolerance.

A reminder that the "struggle" for neuronal survival is not entirely determined by changes in genetic mechanisms but also involves other factors was provided by Dr. A. Baethmann's scholarly review of various *mediator compounds* released following an ischemic insult, which can either enhance or reduce ischemic damage.

The other important consideration, amply demonstrated at this Workshop, refers to the *existence of thresholds*, outlining the ranges for manifestation of certain effects in relation to the density of ischemic insults and associated pathophysiological changes. Thus, it has been shown that an enhanced induction of immediate-early genes, stress proteins, and NtFs takes place within a range defined by thresholds for minimal and maximal density of ischemic insults. The interdependence of thresholds for ischemic injury to density of ischemia was clearly demonstrated in the presentations by Drs. N. Diemer and Wieloch, the former showing the relation of tolerance induction to intensity of insult and the latter demonstrating a threshold for vulnerability of the striatum. With regard to clinical relevance, Dr. U. Ito's presentation showing well-defined thresholds for the induction of ischemic infarction was of a special significance. Using the repetitive ischemia model in gerbils sensitive to unilateral common carotid occlusion, Dr. Ito was able to regularly induce either selective neuronal death or infarction of brain tissue by manipulating the number of repetitive insults and the length of intervals between them. In this model the ability to induce a closely reproducible type of ischemic injury provides an opportunity to study transitional phases between selective neuronal necrosis and infarction, which can contribute to the elucidation of specific mechanisms and events responsible for the development of infarction, a pathological process of great clinical importance.

The role of glia in the pathophysiology of cerebral ischemia and particularly their participation in the maturation phenomenon, was dealt with in several presentations at this Workshop. Being so intimately related to neuronal function, a prompt reactivity of astroglia should be expected, and this was outlined in the presentation of Dr. O. Kempski who described various glial homeostatic mechanisms involved in neuronal protection during the early stages of cerebral ischemia. He pointed out that the typical early swelling of astrocytes is primarily related not to injury but to defensive attempt to reestablish homeostasis, particularly within the extracellular environment. The onset of astrocytic swelling coincides with glutamate uptake and becomes associated with changes in various genetic expressions particularly related to NtFs. Interestingly, as presented in the comprehensive studies of Dr. K.-A. Hossmann, the expression of glial-associated proteins (GFAP, SGP-2) following the ischemic insult occurs appreciably later than neuronal mRNA expressions of immediate-early genes and stress proteins. These differences in the time of appearance, as well as the observation that induction of astroglial-associated proteins begins before the manifestation of neuronal death, indicate that glia play a secondary role in the initiation of events related to ischemic injury. Nevertheless, it becomes evident that glia constitute a potent source for NtFs, which play an important role in protective and regenerative mechanisms after an ischemic insult. Otherwise, the observations presented by Dr. C. Nitsch, demonstrating the persistent upregulation of glial-derived protease nexin 1 in astrocytes, indicate another aspect of astrocytic involvement taking place in the prolonged maturation of ischemic lesions.

In addition to the main topics, additional aspects of maturation phenomenon, such as the role of inflammatory mediators, the effectiveness of delayed postischemic hypothermia, and other factors, were debated at the Round Table discussions. The content and tenor of these discussions emphasized the notion of maturation phenomenon being a battleground for postischemic pathophysiological events which determine the outcome of ischemic insults, as well as stressing the importance and desirability of further studies in this area.

Moderators:
K.-A. Hossmann
I. Klatzo
K. Kogure

Subject Index

Numbers indicate starting page of paper where each subject is included.

GPSR Compliance

The European Union's (EU) General Product Safety Regulation (GPSR) is a set of rules that requires consumer products to be safe and our obligations to ensure this.

If you have any concerns about our products, you can contact us on ProductSafety@springernature.com

In case Publisher is established outside the EU, the EU authorized representative is:

Springer Nature Customer Service Center GmbH
Europaplatz 3
69115 Heidelberg, Germany

Batch number: 10583998

Printed by Printforce, the Netherlands